For the children of Venice – present and future

UMBERTO ALLEMANDI & C.
TORINO ~ LONDRA ~ VENEZIA ~ NEW YORK

Published 2004 by Umberto Allemandi e C.
Via Mancini 8
Turin 10131
Italy
T +39 011 819 91 11
F +39 011 819 30 90

allemandi@allemandi.com
www.allemandi.com

ISBN 88-422 1310-1

Printed in Italy by CAST, Moncalieri - Turin

Edited by Steven Swaby
Designed by Price Watkins, London

Cover: The great flood of 4 November 1966
Endpapers: The lagoon in the 16th century
(after Cristoforo Sabbadino)
Satellite view of the same in the 20th century
Frontspiece: St Mark's Square

Consultant
Deirdre Janson-Smith Science communication consultant
and writer. Graduated in zoology from Edinburgh University.
Trained first in the award-winning exhibition development
department at the Natural History Museum in London.
Consultancy work includes projects with many major
museums, science centres and zoos throughout the world.
Writes on science, medicine and history.

Photographer
Sarah Quill Creator of a unique photographic record of
the city's architecture, environment and daily life, having
worked in Venice for many years. The collection has become a
key resource for publishers and scholars.

**The Venice in Peril Fund thanks the major supporters
of this book and the research project at Cambridge**

The J. Paul Getty Trust
Pizza Express plc
Marina Morrisson Atwater
The Headley Trust
The Boston Chapter of Save Venice Inc.
Sir Mark and Lady Moody-Stuart

Acknowledgements
Appreciation to all the dedicated scientists whose
work has formed the basis of this book, many of whom
generously spent time discussing their work with us.
Dr Tom Spencer (Director, Cambridge Coastal Research Unit),
Dr Pierpaolo Campostrini (Director, CORILA) and
Lady Clarke have provided valuable guidance throughout.
Sincere thanks to Deirdre, Steve, Ray and Ian and staff at
CORILA who have made the book possible.

Contents

Foreword

THE 16 million visitors who come to Venice annually see restoration going on everywhere. This is a city where property values are booming and monuments are expertly guarded by the Superintendencies, the responsible government officials. But the deeper reality is less happy. Time is running out for this loveliest of cities. The frequency of flooding is increasing, and Venice is essentially no better protected from an extreme weather event than it was at the time of the great flood in 1966.

The arguments over whether mobile barriers between the lagoon and the Adriatic are necessary or actually damaging have divided the citizens of Venice and Italian politicians into camps so fiercely opposed that one is reminded of the Montagues and Capulets. The moderate have held back from discussions, not knowing where the truth lies. Vital protective measures have certainly been delayed by at least a decade, and comprehensive long-term planning, which Venice needs to the same degree as The Netherlands, has languished.

The science contained in this book presents a timely distillation of current research and potential solutions. Venice in Peril hopes that those who have the power to take decisions about the future of Venice will feel that they now can do so on the basis of more certain knowledge.

This book reveals that the great majority of scientists believe Venice must have some sort of mobile barrier to protect it from extreme flooding events, but it also makes clear how many other factors are threatening the lagoon, and why the ecologists are right to be seriously worried about its health. Venice cannot be saved without investment in both: barrier and lagoon. To ask people to choose between them is a false dichotomy.

Above all, this book shows that scientists will be crucial to the future of Venice. Italy has for years under-invested in research in all areas of life, and people are beginning to realise how much this threatens the country's place in the world. Science for Venice is one of those areas. If we want our great-grandchildren to see this incomparable achievement of man, we have to accept that there is no final solution to its problems. It will always be work in progress and it will always be expensive – but it is a price worth paying.

Anna Somers Cocks
Chairman, Venice in Peril

Crisis

"Oh Venice! Oh Venice! When thy marble walls
Are level with the waters, there shall be
A cry of nations o'er thy sunken halls,
A loud lament along the sweeping sea!"

Byron, *Ode on Venice*

CENTURIES ago, on each Ascension Day, the Doges of Venice would perform a ceremony called 'the marriage to the sea'. Dropping a consecrated ring into the waters of the Venetian lagoon, they declared, '*Desponsamus te mare*' – 'We wed thee, Sea'. Today, city and sea are still as intimately bound together as ever. But theirs is a marriage in crisis.

The first real wake-up call for Venice came on the fateful night of 3 November, 1966, when a violent storm surge from the Adriatic swept over the city, flooding it to nearly 2m above normal water level in its labyrinthine canals, and left it with no electricity, black oil oozing out of the cisterns, and alleyways strewn with rubbish and the corpses of pigeons and rats. The flood threw a harsh spotlight onto the crumbling city, which was slowly but surely sinking into the waters of the lagoon that gave it life.

THE 'VENICE PROBLEM'

The 'Venice problem' was quickly declared to be of national interest and a priority for action. Italy has since enacted several laws to safeguard the city and lagoon, and a great deal of restoration and protection work against flooding has, and is, being done. This is largely financed by the government (as described in the Appendix). But Venice suffers from far more than flooding, and the task of solving the city's many problems is an extremely difficult one. Scientific and political debates about what to do to protect the city and

left Venetians face the aftermath of the disastrous flood of 4 November 1966

lagoon are also part of its history – the most recent considerations of flood protection began over 30 years ago, culminating in a decision in December 2001 to move forward with an ambitious co-ordinated series of flood protection and restoration projects. Its centrepiece, given a cautious go-ahead in April 2003, is a vast mobile barrier system stretching across the three inlets to the lagoon. Its purpose: to block out the Adriatic Sea during extreme storms, isolating the lagoon from the surging floodwaters.

Opposition to this grand, highly experimental scheme remains fierce, primarily on the grounds that it could destroy the lagoon's already damaged ecosystem, and that it fails to address many other urgent threats to the long-term sustainability of the city. The heat of debate surrounding its planning and development tends to drown out the more complex and subtle arguments about how to protect and restore Venice and its lagoon. But the future of both can never be assured unless the choices made to safeguard them are based on sound science.

DRAWING TOGETHER THE EXPERTISE
In 2001, Venice in Peril funded research at Cambridge University's Coastal Research Unit and at CORILA, the Venice-based consortium for coordinating research concerning the lagoon system. The aim was to bring together reliable research from the many groups and laboratories working in Venice for an international review of the state of knowledge.

The charity, which traditionally funds architectural restoration, cultural and artistic projects, had realised that there would be little point in rescuing individual buildings while the whole city remained under increasing threat from flooding. A much wider perspective was needed.

The aim of both the project and the meeting it inspired (see *An International Conference* on page 11) has been to establish a sound basis of understanding, drawing together the multiple sources of information from many disciplines, often isolated from each other, so that a holistic picture of problems facing Venice – and their possible solutions – could emerge. The ability of experts to share knowledge and experiences is of huge benefit, not only to Venice, but also to other places facing the same issues in managing sensitive coastal wetlands.

Venice was founded in adversity, in the middle of a lagoon, yet this brought extraordinary advantages – from reed huts on stilts, the settlement developed into a maritime superpower and splendid city. The interventions of man throughout Venice's history have focused on maintaining the symbiotic relationship between the city and its lagoon. Now, at the beginning of the 21st century, Venice's survival is threatened by:
above **Fragility** of its crumbling architectural heritage
above, left **Flooding** which has become a chronic affliction
left **Risks** to the efficient running of emergency services

above, right **Ecosystem** degradation and loss of the functions that gave life to Venice
right Scientists are working to address the growing number of challenges to the city's survival

THE NEED FOR INFORMED DIALOGUE

The message from the 2003 Cambridge conference is clear. Most participants agree that the threat of global sea level rise means that Venice's only hope for the future is to be able to block out the Adriatic when the need arises – although the barriers are considered neither a final, nor a sole, solution to the city's woes. Without more well-founded and integrated analyses of the complex scientific issues involved, an appreciation that there is much more to know and a greater willingness to collaborate openly, sharing both problems and solutions, Venice's future looks very uncertain.

Venetians, and the millions of tourists who come to visit this World Heritage city, struggle through winter floodwaters with increasing regularity, past abandoned and decaying buildings. Rising waters and a disappearing ecosystem are accelerating the decline. Global warming threatens its future.

The research effort is immense, with hundreds of experts working on the problem from all over Italy and the world. Their work is vital in helping the many stakeholders to make their decisions about the future of Venice. The chapters that follow aim to summarise the state of scientific and technical research at the start of the 21st century, explaining what is known about the city and lagoon, what is still open to debate and further study, and how scientific expertise can inform policy for Venice in the future. While there is still much more to be learned in the next decades of this century, solutions are required now, and fast.

Summary of conclusions

SETTING

- Venice lagoon is a rich and semi-natural ecosystem, inextricably linked to the life of the historic city and a unique ecosystem within the Mediterranean panorama

- The lagoon emerged 4–6,000 years ago and 1,000 years of human actions and interventions have had a profound effect on its physical structure and ecology

- Today, the lagoon is at risk of becoming a marine bay, due to erosion of sediments and stronger water exchanges with the Adriatic Sea

- As the water deepens, the city is more vulnerable to flooding and associated urban decay

- Vital lagoon habitats have already been lost and those that remain are under threat – and with them, the natural dynamics that the city depends on

FLOODING

- Flooding is caused primarily by storm surges – winds that drive water into the lagoon

- Many man-made changes have inadvertently reduced the lagoon's resistance to incoming waters, against a background of sea level rise – hence higher tides

- It is not a new phenomenon for Venice, but the damaging effects of flooding create an insatiable demand for urban maintenance and are gradually driving the population away

- The idea that Venice is sinking comes from the extraction of groundwater in the mid-20th century, which caused major land subsidence. This has now been stopped and Venice has returned to subsiding naturally and gradually

- Barriers at the inlets will protect Venice from extreme water levels; other measures are being taken to protect the city from medium-high tides and 'chronic' flooding

REMEDIES

■ All projects to restore the lagoon and protect the city must work in a symbiotic way

■ There are two kinds of measures: 'local' measures within the city and the islands, and 'diffuse' measures across the entire lagoon environment

■ Local measures offer vital opportunities to improve the city's infrastructure and restore its buildings

■ Some remedial measures are highly experimental and their effects will only be learned in the long-term

■ Restoring the saltmarshes is key to the future health of the environment, its unique wildlife and the resilience of the entire lagoon system

■ Most scientific authorities are in agreement that the most viable way to protect Venice from extreme flood events is to create a barrier

BARRIER

■ To protect against extreme flooding, construction has begun for the **MOSE** scheme of mobile barriers and breakwaters at the lagoon inlets

■ Further measures are needed to reverse the degradation of the lagoon system, to complement and mitigate the negative impacts of the barrier system

■ Venice can benefit from experience gained in reducing flood risk in other places, notably flexibility during implementation, institutional rigour and unexpected benefits/adaptations of the system

■ This is a new phase for science in Venice, where some old questions are sharpened by implementation of the barriers and many new questions arise

■ The long-term options for Venice will need to be considered in the context of uncertainty over future global environmental change

The Science of Saving Venice

FUTURES

- Human-induced climate change poses one of the biggest long-term threats to the survival of Venice – sea level could rise locally by at least 12cm and as much as 70cm by 2100

- Predicting sea level rise is extremely complex and there is much uncertainty regarding the scale of impact it will have, also depending on adaptation and mitigation measures

- Radical ideas for saving the city include raising the ground level by up to 30cm

- The future health and survival of the city and lagoon depend on creating a balance between the needs of the environment, industry, agriculture, tourism and the Venetians

- Scientists need to work better among themselves and with all other interested parties if sustainable policies are to be developed

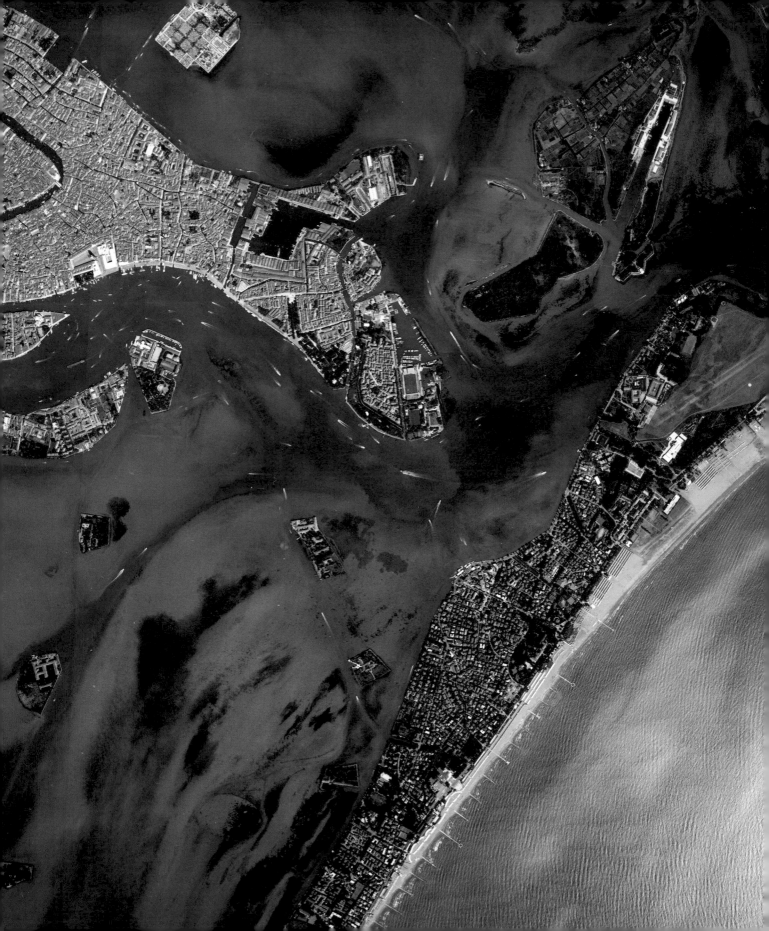

Setting

*"Venice is inconceivable without its lagoon;
it would not, could not, exist without its lagoon."*

UNESCO RAMSAR Report, 2003

VENICE lies in the shallow waters of a coastal lagoon connected to the northern tip of the Adriatic Sea. First occupied in the fifth century, it grew by the 14th century to be the magnificent centre of a major maritime power. Today, it is a World Heritage Site, and the lagoon is recognised as a wetland of international importance. From the beginning, Venetians have managed and modified their watery environment to suit their changing needs, and the lagoon has adapted slowly. But the scale of intervention in the 20th century, and its environmental impact, has been so great that both city and lagoon are now in serious decline. The relationship between man and the environment, city and lagoon, has changed from co-existence to a type of conflict.

THIS CHAPTER EXPLORES
- **THE SETTING OF VENICE IN A MARSHY LAGOON**
- **THE PHYSICAL HABITATS OF THE LAGOON**
- **HISTORIC CHANGES TO THE LAGOON**
- **THEIR IMPACT ON ITS PHYSICAL AND BIOLOGICAL STRUCTURE**

left Seen from above: Venice, lagoon and sea are interdependent

City in the marshes

Venice's past, present and future depends on its relationship with water.
To understand Venice, you must understand the lagoon

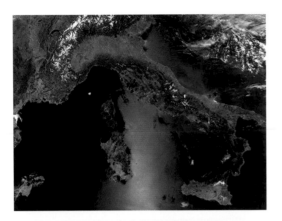

top Venice lagoon lies
at the northern tip
of the Adriatic Sea

above Venice city lies at the
heart of the lagoon – the
canals are its 'veins' and
'arteries', the saltmarshes
its 'lungs'. Large-scale
circulation through the
inlets identifies three
sectors: northern, central
and southern lagoon

VENICE lies at the northern tip of the Adriatic Sea, itself a northerly extension of the Mediterranean. The city is sited at the heart of a tidal lagoon, about 50km long and 20km wide. A strip of land separates lagoon from sea, protected by sea walls and cut by three inlets; several settlements are situated along here. Twice daily, tides flush the lagoon waters, and their ebb and flow shapes both lagoon and city.

17,000 years ago, during the last great ice age, the landscape was very different. The sea was 100 metres lower than it is today and this region was dry land. The seas rose as the ice age ended, and by 4,000 years ago the coast looked much as it does today. Lagoon systems formed along the new coastline as sediments carried down by the rivers built up banks and barriers, and marshes and mud-flats developed in the water trapped behind them. Venice lagoon and the Lagoon of Grado to the north are the only surviving examples; all the others have either silted up or been washed away by the sea.

On such geological timescales lagoons are unstable, constantly evolving systems. Their destiny is to merge with either land or sea. Left to nature, Venice's lagoon would eventually have disappeared. The fact that Venice still exists is due to centuries of human intervention – as we shall see later.

A PRECIOUS ECOSYSTEM

Venice's lagoon is a complex and delicate environment, a place of 'transition' between terrestrial and aqueous environments, fresh and saltwater systems, and its physical form and biology are the result of their interactions. It is fed by freshwater and sediments from the rivers and surrounding land, and also by the ebb and flow of sea water through the inlets. Waves and currents shape its contours: so do the plants that anchor the soft, muddy sediment, and the creatures that feed among them.

Overcome by the spectacular urban architecture, it is easy to overlook the fact that Venice's lagoon is a world-renowned wildlife habitat. It is the largest wetland in Italy and one of the most important coastal ecosystems in the whole of the

above Saltmarsh showing
Salicornia veneta, a species
which is unique to the
Venice lagoon

far, right The lagoon is an
important sanctuary for
wetland birds such as
these black-winged stilts

below, right Traditional
fishing in the lagoon using
an ancient technique. In
the 19th century, 80–90
percent of fish were still
caught this way. Use of
this system is falling fast
from over 100 fishermen
five years ago to less
than 70 today

Mediterranean. Tides rise and fall far more than elsewhere in the Mediterranean, creating a unique ecosystem that supports a rich biodiversity – some plant species exist only here; a few types of bird and fish depend on Venice for key stages in their lifecycle and many species common in the lagoon are rare elsewhere in the Mediterranean. Many important waterbirds, such as the redshank and the Sandwich tern, breed here. And more birds overwinter here than anywhere else in Italy – thousands of dunlin, for example, stop off for the winter to feed on the mudflats. The lagoon is actually a fine mosaic of different habitats, each shaped by local conditions – the salinity of the water, or whether it is calm and shallow or subject to fast-flowing currents.

UNDERSTANDING THE COMPLEXITY

This is ostensibly one of the most studied areas of the world, due to its compelling and longstanding integration of man and nature, its unique biodiversity, as well as the government resources and public institutions' attention to safeguarding Venice since the latter 20th century. However, there is still much to learn: advances in supporting scientific theory, experimental techniques, physical, chemical and biological monitoring programmes are improving our understanding of the phenomena. Meanwhile, the lagoon is changing faster than ever before and human intervention is on an unprecedented scale. It is a difficult challenge to build up a solid knowledge against which to interpret the impacts, past and future, of human activities. Much work has and is being done. Much more has yet to be done.

The changes in the lagoon are a result of a very complex mix of natural processes and human intervention. To really grasp today's challenges for Venice, we need to be able to understand the interaction between the city, the physical processes that take place in the lagoon and Adriatic Sea, and the chemical and biological 'metabolism' of the lagoon. Scientists observe and monitor the changes in the lagoon in order to understand the processes occurring, but separating the cause from the effect in such a dynamic and complex system is difficult.

"The lagoon is one of the most studied environments in the Mediterranean, yet research is still needed to gain an integrated and deeper understanding of its dynamics"

Pierluigi Viaroli, Dept. of Environmental Sciences, Università di Parma

Profile of the lagoon

VENICE's lagoon is very shallow for most of its area: the average water depth is only about one metre. Natural creeks and channels wind across its bed, carrying tidal waters between sandbanks, mudflats and saltmarshes. Artificial navigation channels cut deep into the lagoon bed, some as deep as 20 metres. There are four main physical habitats, defined in relation to the tides. Some 60 percent of the lagoon lies permanently under water; 25 percent is periodically exposed to varying degrees by falling tides (marshlands and mudflats). The rest consists of islands, in principle always above water, although increasingly vulnerable to flooding.

A **Under water** Open waters and shallows including the natural creeks and dug channels that cut across the lagoon. The deeper navigation channels are marked out by distinctive trios of wooden stakes. Seagrasses grow in the shallows. They help to stabilise the lagoon-bed and are a nursery for fish reproduction

B **Mudflats** Low-lying areas exposed only at low tide. They drain off the minor channels (tidal creeks) and influence saltmarsh accretion and erosion processes. Although they may look uninviting, they are rich in invertebrate life (including worms and clams – the latter are economically significant) and are important as feeding grounds for birds

C **Saltmarshes** Higher-level areas, partially covered with water only at high tide. They are irregularly distributed around the lagoon and range in size from a few square metres to several hectares. Their salt-tolerant plant communities support a rich and diverse wildlife

D **Islands** These areas are not normally affected by high tides: the islands (including Venice), islets and the three strips of land separating the lagoon from the Adriatic (Pellestrina, Lido and Cavallino), fronted by a sea wall

E **Drainage basin**

F **Fish farm**

G **Sea wall**

H **Adriatic Sea**

The Science of Saving Venice

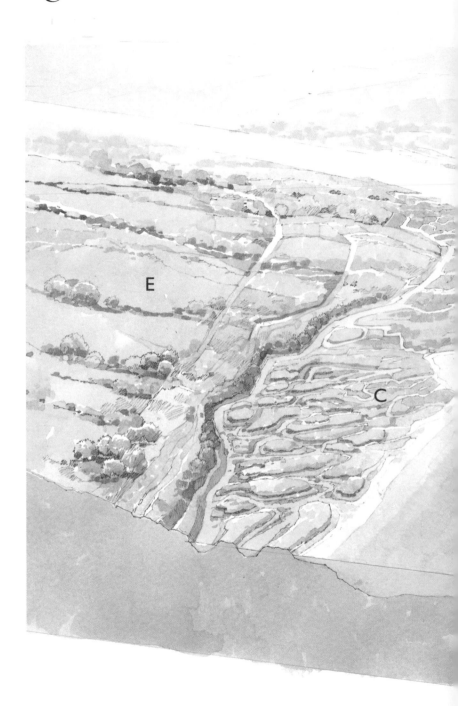

This illustration does not represent a real view, but brings together the main habitats and surroundings of the lagoon

A history of intervention

Ever since people have lived in the lagoon, they have altered it to suit their needs.
Now the scale of human activity threatens to destroy it

VENICE was founded in the fifth century, after peasants fleeing from the barbarians found shelter on the small marshland islands of the lagoon. Over the following centuries, the city grew in size and importance. The original cluster of 118 islands was built upon and gradually joined together, linked by bridges, connected by canals. More marshland was drained. By the 14th century, Venice was the centre of a wealthy and powerful European trading empire.

Venetians depended on their watery environment for food, transport and protection – the shallow marshy waters of the lagoon were famously 'too shallow for invading ships, too deep for marching armies'. The lagoon itself was continually refreshed by the tides, purifying and oxygenating the water. The lagoon waters flushed the canal thoroughfares twice daily and carried away the sewage, which became nutrient for the lagoon's plant and animal life. Venetians harvested the shellfish and fish, and hunted the wildfowl. Many of the traditional techniques persist to this day and fishing still support the local economy.

CHANGING THE LAGOON

Underlying this intimate relationship between citizens and lagoon was a kind of compromise between natural forces and human needs. Natural processes, working over geological timescales, were gently shifting the lagoon to land, as the sediments brought by the many large rivers that fed into it accumulated. But the growing city depended on canal navigation for the trade that brought it wealth and power, and for its protection. So a remarkably bold series of modifications began in the 15th and 16th centuries to keep the lagoon under water.

Seven river mouths were diverted along canals to north and south. Over the next couple of centuries, the diversions were adjusted further as the problem of silting up shifted to new areas, or land began to erode.

The last great works of the Venetian Republic (which ended in 1797) were the *conterminazione lagunare*, which officially fixed the lagoon boundaries, and built the *murazzi*, great sea walls designed to keep the Adriatic out. From the mid-19th to early-20th century, long jetties were built at each of the inlets to accentuate the tidal current which kept sediments from clogging up new navigation channels.

above, left In the 16th century, the main rivers were already being diverted from discharging into the central area of the lagoon

above The 20th century saw the growth of industry and other activites around the lagoon, including the airport (left of picture)

"Too shallow for invading ships, too deep for marching armies"

Paths of the lagoon's main rivers in the 14th century and the less regular inlet system

By the 20th century, many of the rivers had been diverted, via canals, to north and south

INDUSTRY ARRIVES

In the 20th century, the pace of change accelerated. Industry arrived at Marghera on the adjacent mainland and the port was moved nearby. Large areas of marshland were drained for the industrial zones, airport and agriculture. The outer reaches of the lagoon were closed off to protect fish-farms, and new islands were created. Much deeper navigation channels were dug to bring far larger ships – giant oil tankers and container ships – into the new port and industrial zone.

All these interventions have had major long-term consequences. The natural extent of the lagoon has shrunk by over 20 percent, and its physical structure and hydrodynamics (water flow) has been changed for the worse: habitats such as saltmarshes have been lost and water flows are now greater, which increases erosion. In particular, scientists have identified the deep navigation channels as major culprits in causing erosion and distorting the pattern of the tidal currents and trapping sediments, taking them away from the inter-tidal areas.

KEY CHANGES TO THE LAGOON

C13th–14th	coastal protection – tree felling banned in coastal zone and chalk boulders positioned to strengthen sea defences
C15th–17th	diversion of all large rivers discharging into the lagoon
C18th	construction of *murazzi*, sea wall defences
C19th–20th	construction of jetties at the three inlets
mid-C20th	development of Marghera industrial area
	reclamation of saltmarsh for agriculture land
	closing off of fish farms
	construction of two major navigation channels

Lagoon to sea

Venice's lagoon is becoming deeper and more like a sea bay. Its physical structure and ecology are deteriorating, making the city increasingly vulnerable to storm surges and causing species loss in the lagoon

MAPS, surveys and other historical evidence all show that the lagoon is progressively evolving towards a marine landscape. The waters are becoming deeper, marshes and mudflats are disappearing, erosion increasing, and the water has become as salty as the sea.

It is difficult to quantify the change accurately, owing to the complexity of the processes and in some cases to a lack of data, but the problem arises from a shift in the balance of sediments entering and leaving the lagoon.

SEDIMENT LOSS

Sediment characteristics underpin many of the lagoon's physical habitats and determine their health status. Sediments range from fine mud and fluvial silts to coarser sand and their behaviour varies accordingly. They are stirred up and redistributed by currents, deposited again elsewhere or swept out to sea on the tides. But while sediments are accumulating in some areas of the northern lagoon, in others they are eroding and overall, a great deal of sediment is disappearing from the lagoon system. So the dynamics are extremely complex. Measuring the lagoon's 'sediment budget' – the material gains and losses over time – is essential to determining the fate of the lagoon in the modern era. Yet it remains a hugely difficult task. Although while most experts believe there is an annual net loss to the sea, estimates vary as much as ten-fold.

Sediment loss has been identified as a key factor in the process of evolution from lagoon to bay. Diverting the rivers to prevent Venice from silting up has created the opposite problem – starving the lagoon of this sediment source. The long jetties designed to stop the inlets silting up with sediments have played their part in the sediment deficit. General deepening of the lagoon (explained in the next chapter) and deeper canals make currents stronger. Changes in current flows also mean that a great deal of silt and sand are drawn into the navigation channels from the adjacent shallows, which causes further erosion and has to be removed by costly dredging. The total volume of water in the lagoon is said to have doubled in the past century, notwithstanding the reduction in total area, mentioned above.

below, left
Lagoon to bay
The bay is deepening, and the lagoon bed is being smoothed out. With deeper water and a loss of the dendritic structure (creeks) and physical habitats that dampen down the effect of waves and currents, the tides and surges have a more direct impact on water levels in the city and lagoon

below Cross-section of changes to the lagoon bed due to erosion – an 'underwater desert' in some areas

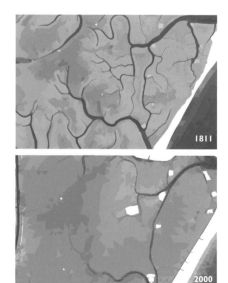

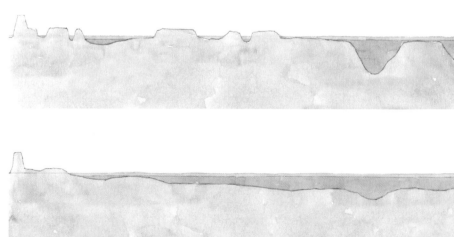

CLAM FISHING AND SEAGRASSES

CLAM fishing using large mechanical equipment (below) like this, causes long-term damage to the lagoon bed by stirring the sediments and interfering with the lifecycles of animals living in them. The cloudy water prevents essential sunlight from reaching plants on the lagoon bed and the intensive harvesting of clams dramatically reduces the population's revival rate. The interference with seagrass communities is particularly significant, as their long roots are needed to anchor the soft sediment (right). A major census of the seagrass communities, undertaken in 1990 and 2002, has yielded a detailed map of species distribution. A small general loss across the lagoon over the previous ten years is actually composed of larger losses of species in some areas contrasted with their reappearance in others. This study demonstrates the difficulty of making generalisations about what is happening in the lagoon.

right **Malamocco inlet** Mathematical modelling reveals how the jetties at the inlets influence water flows in and out of the lagoon, which in turn govern the sediment transport

A Early 1800s
Water flows exit and return through the inlet with the ebb and flow of the tide, in apparently equal quantities

B Now
Jetties and associated coastal structures have altered the flow of water, so that sediments swept out on the ebb tide are carried further along by the coastal current, which limits the amount of sediment returning to the lagoon on the subsequent flow tide

Incoming (flood) tide   **Outgoing (ebb) tide**

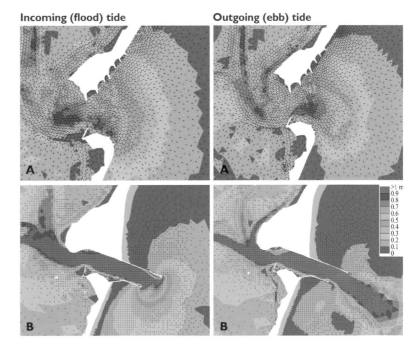

Ecological impacts

Venice's saltmarshes and mudflats are disappearing, starved of sediment
and eroded by waves as the lagoon deepens

SCIENTIFIC surveys show that saltmarshes and mudflats are down to a third of their extent (about 47km²) at the end of the 19th century. Around 20 percent of the lagoon's plant and 50 percent of its bird species have been lost since 1930. Species that live on the bottom of the lagoon were particularly affected, as much of the intricate sub-structure of winding creeks and meandering canals within the shallows have been destroyed by stronger currents and wave erosion. This lagoon 'physiology' (habitats and structures) is vital to slowing the flow of currents in the lagoon, and their irregular form moderates wave energy. They are also critical to the chemical and biological functioning of the system.

Tidal creeks wind their way through a maze of tiny 'islands' of salt-tolerant vegetation on the marshes, specially adapted to the alternating wet and dry phases. As tidal creeks and natural channels disappear, 'ponds' within the marshes are expanding and joining up with the open water and further fragmenting the marsh structures.

Some areas of the lagoon are faring much better than others. The processes mentioned above have removed nearly all saltmarshes from the central areas. In the south, original reedbed marshes have transformed into saltmarshes since the river diversions, but the process is all but stable. Only the northern lagoon retains some original characteristics and scientists are working hard to determine exactly what factors sustain them, so that more effective work can be done to revive and protect the rest.

Looking to the future, many lagoon specialists are concerned that with the lack of sufficient sediments and rising water levels, the saltmarshes and mudflats will continue to degrade and possibly disappear within a century. Any sea level rise due to human-induced climate change will of course exacerbate this trend.

POLLUTION HAZARDS
In addition to physical changes to the system, and the effects on saltmarshes and their associated ecosystems, the lagoon was exposed to major pollution since the mid-20th century, with some dramatic consequences for the environment as water quality deteriorated.

below, left
Cruise ships enter through the Lido inlet, pass through the Guidecca Canal and moor in Venice's passenger terminal

below, right
Industrial activity has left areas of land adjacent to the lagoon that are highly contaminated, and large areas of Marghera now lie abandoned

Venice's lagoon is fed by a large drainage basin – a 1,850-km² area – and despite the redivertion of many key rivers, there are still some river systems and associated canals that flow into the lagoon. They bring with them toxic metals, pesticide residues and nutrients from industry, agriculture and urban areas.

In the past, the lagoon trusted primarily in tidal exchanges with the Adriatic and chemical breakdown by micro-organisms to cope with this influx. Micro-organisms in the mud and water broke down organic matter and waste water was diluted by the daily tides. But the input, since the last century, of persistent and potentially toxic metal and synthetic compounds, as well as greater quantities of other contaminants, has meant that the lagoon's natural waste treatment system can no longer cope. Control of industrial waste and agricultural run-off started in earnest in the 1990s, and a major integrated waste-water treatment plant is currently being built. But there is still an urgent need for the sewage inputs to the lagoon and canals of Venice to be more strategically addressed.

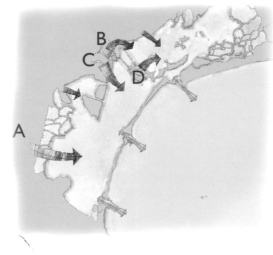

Main sources of pollution
- Agricultural run-off: pesticides and fertilisers
- Industrial waste
- Urban waste: sewage etc from Venice and mainland
- Air pollution: air and road links, industry

A Drainage basin
B Mestre
C Port and Marghera
D Venice

Ecological impacts continued

The Science of Saving Venice

ALGAL BLOOMS

Nitrogen and phosphorus compounds are vital nutrients for a healthy ecosystem. But if there is too much, the system suffers. A process called eutrophication begins – the nutrients fuel explosions of growth in certain algae, robbing the water of oxygen and starving out other species. In the 1980s, Venice lagoon suffered terribly from dense algal blooms every spring, clouding the water and suffocating life on the lagoon bed. Populations of a larger invasive seaweed called *Ulva rigida* also exploded, and soon accounted for some 80 percent of vegetation growth. It had to be harvested regularly to stop it choking the lagoon.

Tough new regulations on industry and agriculture finally brought the nutrient overload under control in the 1990s, but although phosphates and ammonium levels have reduced, high quantities of nitrates still persist. It still isn't fully understood why the seaweed explosions died away, so logically it remains possible that they may return.

TOXIC TIME BOMB?

Venice is one of Italy's main ports, and major chemical/petrochemical industries are still present at Marghera. Both the port and the industrial complex were allowed to develop unchecked from the 1950s, and used to discharge their waste directly into the lagoon. As well as the catastrophic risk of oil-spills or chemical explosions in the lagoon, a longer-term threat exists.

Toxic metals (such as mercury, copper, arsenic and lead) and persistent substances such as organochlorine pesticides and dioxins have accumulated in the lagoon sediment and canals, storing up problems for the future. Uncontrolled dumpsites in Marghera, created in the 1970s, are still so toxic that they must be capped and isolated from the lagoon system. Erosion and dredging stirs up the sediment, sending pollutants back into the system. Our understanding of the effect these contaminants have on individual species is improving, but their impact on whole ecosystems is now a major area of scientific study. In addition,

Effect of excess nutrients

In a healthy system, seagrasses grow on the lagoon bed, helping to anchor the sediment and moderate wave action. Sunlight and oxygen penetrate the shallow water and support their growth through photosynthesis

When excess nutrients enter the system, algae reproduce rapidly causing 'blooms' that consume available oxygen dissolved in the water. Tougher seaweeds like *Ulva rigida* thrive. Less vigorous seagrasses are smothered by these and, cut off from sunlight, they are unable to photosynthesise, and so die

As seagrass vegetation dies off, there is nothing left to anchor the sediment. It is swept up by currents and clouds the water, further decreasing sunlight penetration. The lagoon bed becomes smooth and no longer dampens wave actions. Erosion therefore increases in a feedback loop, making it increasingly hard for seagrass recolonisation

above Wave energy (*moto ondoso*), which increases as the water gets deeper in the lagoon, is a key factor in changing the physical and ecological structure of the lagoon. In more open parts of the lagoon, saltmarsh edges are crumbling and vegetation is dying off. Higher water levels also mean that the marshes are underwater for longer, which limits the viability of certain species

above, right Accumulation of 'slime' created by an algal bloom where there normally would be a pleasant lagoon beach at the island of Sant'Erasmo

as analytical methods improve and more contaminants are measured, new threats to the system (e.g. endocrine disrupters) are emerging.

Preserving the lagoon's unique ecology, while still maintaining the industry and port and other productive activities, has been, and will continue to be a major challenge.

EXPOSED TO THE TIDES

Long-term studies confirm that the primary threat to the lagoon comes from its *physical* degradation – which strips away important habitats. Where there was once a complex mosaic of habitats, there is increasingly a scoured flat bay. The sediment deficit, the scouring of the navigation channels, and the decline of the biological communities, have all led to the loss of key physical features and habitats, and deepening of the lagoon.

But deepening is also the result of global sea level rise and the natural and human-induced sinking of the land. In the 20th century, Venice lost some 23cm in height relative to the water level. As the waters deepen and the tides' impact grows, the problems of flooding increase. This is an immediate threat that must be confronted.

CHAPTER 2 SUMMARY

- Venice lagoon is a rich and semi-natural ecosystem, inextricably linked to the life of the historic city and a unique ecosystem within the Mediterranean panorama
- The lagoon emerged 4–6,000 years ago and 1,000 years of human actions and interventions have had a profound effect on its physical structure and ecology
- Today, the lagoon is at risk of becoming a marine bay, due to erosion of sediments and stronger water exchanges with the Adriatic Sea
- As the water deepens, the city is more vulnerable to flooding and associated urban decay
- Vital lagoon habitats have already been lost and those that remain are under threat – and with them, the natural dynamics that the city depends on

Flooding

"No degree of flooding is acceptable
in a city of Venice's supreme importance"

Anna Somers Cocks, Chairman, Venice in Peril

OF all Venice's problems, flooding is the most obvious and immediately alarming – water overflows the embankments and seeps up through the drains. There are two types of *acqua alta* (literally 'high water'): fairly routine inundation of the lowest areas of the city and rare but devastating extreme events. While *acqua alta* has been a part of Venice's history since its foundation, flooding is on the increase, and it is doing great damage to the city, both physically and socio-economically. More flooding is the result of a combination of natural forces (winter storms, combined with long-term processes of rising sea level and subsiding land) and man-made changes to the lagoon that have reduced its natural defences and caused the land to subside.

THIS CHAPTER EXPLORES
■ **WHY AND WHEN VENICE FLOODS**
■ **WHY THE CITY IS FLOODING MORE FREQUENTLY**
■ **HOW HIGH WATERS HARM THE CITY AND THE LIVES**
  OF ITS PEOPLE

left Venetians are becoming used to regular flooding of their city

Why does Venice flood?

Venice floods for many reasons: it is subject to storm surges, the land is sinking, global sea level is rising, and changes to the lagoon and its inlets have reduced its capacity to moderate water levels in the city

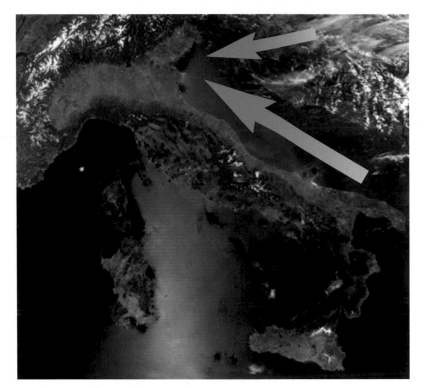

VENICE was built at sea level. Twice a day, the level of water in the city rises and falls as the tide ebbs and flows through the inlets that connect the lagoon to the Adriatic Sea. The tidal range of over 1m here is greater than at the southern end of the Adriatic, where tides usually vary by a few centimetres. But tides alone are not the reason that Venice floods. The main influence comes from low pressure weather systems and associated winds, which create storm surges that effectively push more water into the lagoon.

Two wind systems in particular create these surges. The *Bora* from the north-east acts locally to drive a high surge of water across the lagoon – sometimes creating a difference of up to 20cm in water level between northern and southern parts of the lagoon. The south-easterly *Sirocco* wind, from the Sahara, forces surges up the long narrow 'funnel' of the Adriatic Sea, towards the lagoon. Water rushes through the three inlets and sweeps across it to the city – arriving at the Grand Canal about 45 minutes later.

It was a very unusual combination of conditions – low pressure, persistent rain and high winds – that caused the tragic flood of 1966, which took water levels in Venice nearly two metres higher than usual. Most people believe it is now only a matter of time until the next disaster and there have been several incidents of grave (but not catastrophic) flooding since then.

Flooding remains one of Venice's most serious issues not just because the 1966 disaster focused attention on the vulnerability of the city's precious heritage and residents, but also because moderate (chronic) flooding is occurring increasingly frequently. This is because Venice has sunk. As it is lower in relation to water level, lesser tidal surges can cause flooding – and, as described in the previous chapter, as the lagoon becomes more like a marine bay it has lost some physical features that used to offer a natural dampening effect on incoming waters, so more water comes into the city with each tide and surge.

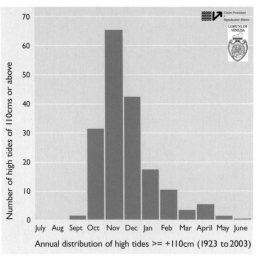

Annual distribution of high tides >= +110cm (1923 to 2003)

Flooding is largely confined to the winter months

Main winds targeting Venice
Responsible for storm surges in the Adriatic:
Bora – which blows from the north-east
Sirocco – which blows right across the Mediterranean from the south-east and up into the 'cul-de-sac' Adriatic

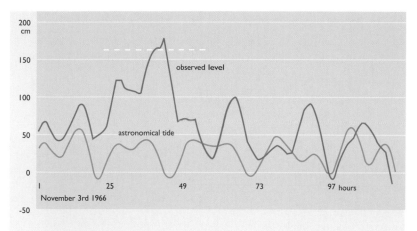

200 cm

150

100

50

0

-50

observed level

astronomical tide

1   25   49   73   97 hours

November 3rd 1966

**Storm surge of
4 November, 1966**
During the main surge, waters rose more than 1.9m above reference sea level, and beyond the instrumental measuring limit of the time. It was followed by a series of 'seiches', lower but still high tidal surges that are caused by the water oscillating in the blind end of the northern Adriatic basin, and which mean that severe storms can have effects for days.

The dotted line shows the 'astronomical' tide rhythm – the predictable rise and fall under the influence of the moon – and highlights the importance of weather features in determining water level in Venice

MEASURING SEA LEVEL

LIVING so intimately with the sea, Venetians have always been careful observers of water levels in the city. Sea levels in Venice have always varied widely, not just over hours and months, but over longer timescales, which can make it difficult to deduce an emerging pattern or trend. Historical water levels have been estimated using archaeological data from Roman times to the 13th century – items excavated at differing levels are interpreted in terms of their assumed function, so a boat relic probably marks water level whereas household objects would have been situated a little higher than water levels in canals. From about the 15th century, buildings were often engraved with a 'C' to show average high tide levels. This level is also marked naturally by a green line of algal growth and some ingenious extrapolation has been done by observing representations of the green line in 18th-century paintings. Instrumental measurements began in Venice in the 1870s, and in 1897 a fixed reference point was established at Punta della Salute, at the entry to the Grand Canal. Like the Venetians themselves, this reference point is used for expressing water level throughout the book.

Average water level in Venice (commonly but inaccurately referred to as average 'sea level') is calculated as a mean annual value, taken by averaging all the tidal maxima and minima for a whole year. But it is important to bear in mind that today's mean water level is about 25cm above the Punta Salute reference zero – the combined result of sea level rise and land subsidence. The lowest parts of the city now lie less than half a metre above today's mean water level (70cm above the 1897 reference level) and are therefore vulnerable to tides only a little above average.

Punta della Salute is the reference tide gauge station on the Grand Canal, in front of the old customs house

Normal tide range is between -50cm to +80cm, with respect to the Punta Salute 'zero' reference. A 'sustained' tide is above 80cm and an 'exceptional' event occurs when water level reaches 110cm or above, and the Town Council's flood warning system is activated. (This is what is called 'chronic' flooding.) A level of at least 140cm defines an 'extreme event'. Sometimes several flooding events occur over a number of days, as occurred in 1966, due to seiches.

Forecasting sea level
There is an extended network of tide gauges around Venice and the lagoon which, in turn, are integrated with other sea level monitoring networks and meteorological stations across the whole region. All this information is analysed (using mathematical models and statistical analysis) to interpret past and present trends as well as to forecast water levels and flooding.

Sinking city, rising seas

Natural processes are causing Venice to sink gradually, while the seas are rising around it.
But human actions greatly accelerated the process in the 20th century

SINKING LAND

IT was inevitable that Venice would sink. The land itself is subsiding naturally as the ancient sediments of the coast settle. Also, the movement of the Earth's crust (on a geological timescale), is driving this part of Italy down under the Alps. Together these processes cause Venice to sink by about 0.5mm each year, although the exact amount varies over time, from place to place – the northern and southern parts of the lagoon have registered rates of subsidence of a few mm/year.

That the city has not sunk more is thanks to a layer of highly stable, consolidated clay, known as *caranto*, which lies several metres below the surface. This clay layer formed about 6,000–10,000 years ago as the seas, which once covered this region, retreated. Venice's foundations consist of wooden stakes anchored in this layer and the natural irregularity of the *caranto* is one of the reasons why ground levels vary around Venice.

Archaeological studies have calculated a loss in height relative to sea level in the order of 1–1.5mm a year. Architectural studies on the Basilica of St Mark's confirm this, revealing an estimated reduction in the margin between ground level relative to sea level of about 14cm per century.

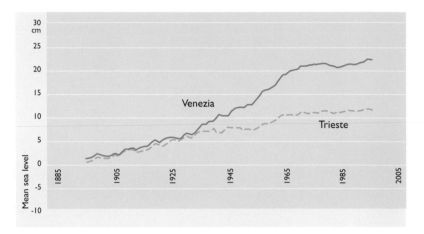

above Sea level rise in Venice and Trieste, separated by 100km in the northern Adriatic. The rate of sea level rise was the same in both places until the striking acceleration in Venice, from the mid-1920s. Today, the rates are back in line with each other (dotted line), although relative sea level in Venice is now higher due to the subsidence caused by groundwater extraction

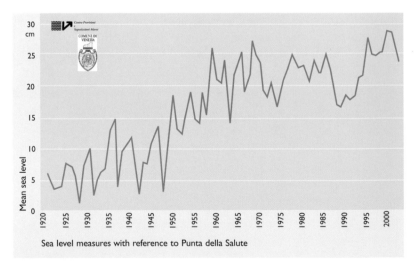

Sea level measures with reference to Punta della Salute

Rising average sea level in Venice: the general upward trend is clear although variations from year to year are significant

GROUNDWATER EXTRACTION

The rate of land subsidence in the lagoon accelerated dangerously in the last century. From the 1920s, industries started to settle on the mainland perimeter in the area known as Marghera, and the port of Venice was also relocated here. The huge volumes of water needed to support these activities were pumped out of deep aquifers (natural underground reservoirs) beneath the lagoon. Unfortunately, it was not realised for decades that these aquifers were an important 'cushion', buoying up the land on which Venice rested. Land began to subside dramatically, as can be seen from a comparison with measurements at Trieste, which is subject to the same 'background' subsidence and sea level rise.

In less than 50 years, from the 1920s to 1970, Venice had sunk by approximately 10cm more than Trieste. Alerted by geotechnical specialists, who attributed this dramatic drop to groundwater pumping in 1970, the government banned it and the subsidence rate adjusted back to normal within a few years as the aquifers refilled. However, the damage had largely been done in terms of Venice's protective margin against flooding and the land rebound was relatively small (2cm) as the basal clay on which the city rests had been irreversibly compacted.

CITY ON STILTS

AS Venice moved from temporary refuge to permanent city in the ninth century, people found a radical solution to building on the unstable marshy land. They drove wooden piles – long, sharpened poles of alder, oak and larch – into the more stable subsoil of clay under the lagoon. The airless muddy soil preserves the wood and stabilises foundations (partly thanks to friction).

An astonishing number of pilings were used: the Rialto Bridge is built on 10,000, the Chiesa della Salute on the Grand Canal more than a million.

A layer of oak was laid across the piles to form a foundation. Above this was placed a layer of Istrian stone, a tough marble from the mainland that is impermeable to water. This provided a protective barrier against the lagoon waters.

Above the water line, lighter-weight materials – brick, wood and plaster – were used. This reduces the compaction effects of the buildings and confers flexibility as the various ground layers shift continually.

Sinking city, rising seas continued

CLUES FROM CANALETTO

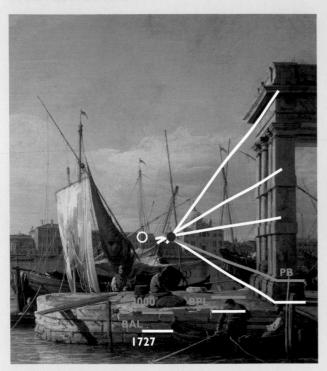

CANALETTO's meticulous paintings of Venice in the 18th century provide a unique historic record of the city's relationship to the water, because of their accuracy. Canaletto and his student and nephew Bellotto used a 'camera obscura' to paint their scenes. This device reflects a mirror image of the view onto paper, which can then be traced on – the picture is in essence a photograph. Modern photographs from the same viewpoint have been used to compare water levels – and reveal where ground levels have changed.

The Punta della Dogana example shown here illustrates the difference clearly. The bases of the columns to the right show that the whole pavement has been raised, and still the water is higher up. The rise was calculated as 70–110mm over two centuries.

RISING SEAS

Subsidence of the city has been accompanied by a gradual worldwide sea level rise. In line with rising global average temperature, seawater volume expands and land-based ice melts, raising sea levels across the world. This global process is known as 'eustasy'. Concerns for Venice's future are in part based on predictions of a global rise of about 8–88cm by the year 2100, but the regional implications have yet to be worked out. At least for the time being, sea level rise in the Adriatic region seems to have slowed down (Chapter 6 explores this in more detail).

Although it is the extreme events that Venice fears most, it must also cope with and plan for a general and gradual rise in water levels. Since the 1966 flood, people have moved from vulnerable ground-floor homes, but stores and workshops remain at risk. The city's highest land is no more than 2m above sea level, and much of the city lies much lower than that. Areas that were once safe, now flood, and low-lying areas are flooding with increasing frequency.

> *"Venice has lost a century in its battle with the sea"*
>
> Prof Albert Ammerman, archaeologist, Colgate University, USA

Turn of the 20th century

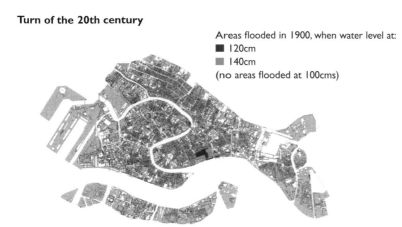

Areas flooded in 1900, when water level at:
- ■ 120cm
- ■ 140cm

(no areas flooded at 100cms)

Today

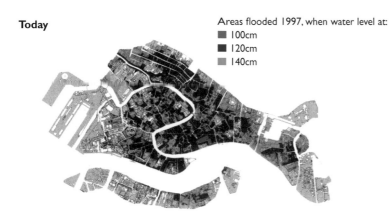

Areas flooded 1997, when water level at:
- ■ 100cm
- ■ 120cm
- ■ 140cm

These two maps show that much more of the city floods today when water levels rise than a century ago

FLOODING IN VENICE TODAY

Water level (cm)	Area of city flooded (%)
25	0 (today's mean sea level)
70	St Mark's Square and Basilica begin to flood
100	3.5
110	12
120	35
130	69
140	90

Flooding more frequently

Flooding in Venice is occurring more often. In winter, what was once an occasional event is now an almost daily occurrence in some parts of the city

THE first record of *acqua alta* dates back to 1240, when waters were 'above man height' in the streets. It is something people have learned to live with but, increasingly, the dampness and complications caused in daily life are becoming unbearable. Today, in winter, raised walkways (*passerelle*) are out most of the time in the low-lying areas and along strategic routes, such as from the railway station. Water wells up from drains to pool in the city's squares and court-yards, and may flow over the canal banks. Many entrances to buildings, schools, shops, store-rooms and church crypts flood, while most ground-floor dwellings have been abandoned. Even emergency services are jeopardised by high waters when water level in the canals leaves too little space under bridges for the fireboats and ambulances.

The present situation is significantly worse compared with 100 years ago. Of the ten highest tides (above 140cm) recorded in the century to 2002, eight occurred after 1960. In the winter of 2002 alone, in the space of three weeks (14 November–8 December) there were ten flood events above 110cm, five above 120cm and one above 140cm. But the picture is complex due to irregularity within the general trend. In the whole of 2003, water level did not rise above 110cm. Only by comparing data over many years can meteorologists be sure that the frequency of flooding is not just linked to some cyclical trend but a permanent change.

Scientists have linked increased flooding to the rise in average sea level attributable mainly to the episode of human induced subsidence, and to changes in the lagoon's physical structure, which affects the entry and movement of water within it. Changes in the shape of the inlets, loss of salt-marshes and deep navigation channels have contributed to a greater volume of water and stronger current entering the lagoon when pushed by the tide and winds. A small effect is also attributed to the reduction in total area of the lagoon, due to land reclamation and other inter-

ventions, so the greater volume of incoming water has less area in which to spread itself.

Even when the tide is not particularly strong, high water regularly tops the protective layer of Istrian stone on the canalside buildings to eat away at the plaster and brickwork physically (wave and propeller energy) and chemically (salts corrosion). The city is under siege.

"Venetians have been getting their feet wet for centuries"

Paolo Costa,
Mayor of Venice

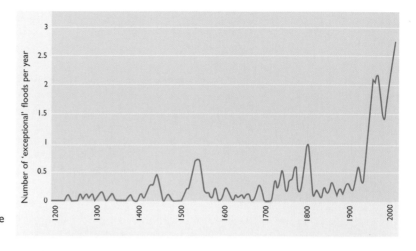

Flooding since 13th century, partly based on interpretation of historical accounts of the life and times of Venice

Since 1923 there has been an increasing frequency of 'normal' flooding events (over 80cm) – when less than three percent of the city is inundated

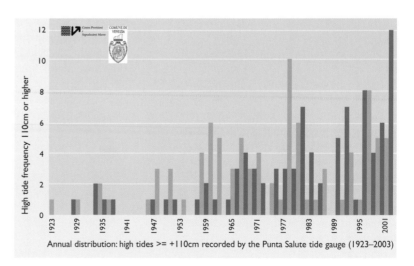

Annual distribution of high tides >= +80cm recorded by the Punta Salute tide gauge (1923–2003)

Analysis of the past 30 years' data shows fewer than four such 'extreme' episodes per year

Annual distribution: high tides >= +110cm recorded by the Punta Salute tide gauge (1923–2003)

City under attack

Venice's world-famous architectural heritage demands constant attention. High waters, waves generated by boats and pollution all attack its fabric and foundations

VENICE's architectural problems are not new. The battle to maintain the delicate fabric of the city has been relentless. Up until the 19th century, canals were filled in, buildings demolished and rebuilt in the face of changing water levels, but Venice's '*forma urbis*' and architectural heritage are now under strong preservation orders. Today, the city's extreme fragility is startlingly obvious – especially the many crumbling buildings that have been abandoned as a result of the collapse in the local population and the relocation of many offices and businesses to the mainland.

The main decay is caused by the destructive action of sea water. Since the tidal exchange between the lagoon and Adriatic has become stronger, salinity of the canal water is the same as the sea. On many buildings, water level now comes above the impermeable Istrian stone base for most of the time, soaking into the porous brickwork, wood and plaster. Dissolved salts degrade the building fabric as they crystallise within the walls. Air pollution used to be a problem and corroded the marble facades of Venice's palaces. But Venice was one of the first Italian cities to convert to methane power generation (instead of coal and oil) and industrial emissions are being controlled.

TRAFFIC WAVES

Buildings are also challenged by waves set up by speeding boats carrying people and goods through the city's canals. The intensity of the waves has been estimated as adding a few centimetres to the mean water level. While the waves attack surfaces above water, the action of propellers below water creates a sucking force, eroding cement and foundations. Stirred up sediment clogs the canals and obstructs the ancient sewage system on which Venice still relies.

The impact on the people of Venice is enormous: the inconvenience of flooding, and the ongoing damage to their fragile buildings, and the overwhelming presence of tourists have meant that many Venetians have abandoned the historical centre as a place to live or work. In the 1950s the city boasted a population of 150,000. Today, it is closer to 65,000. If future generations are to be drawn back to the city, remedies are urgently needed to mitigate flooding and urban degradation.

left Fast and intense boat traffic causes damage – by setting up higher waves, and 'sucking' at the foundations. Studies are still trying to assess the consequences of passing cruise ships for buildings and air quality along their route through Venice

above Cross-section through building showing structural damage. Motor boat engine propellers cause scouring of the mortars and brickwork below water. Wave energy and the infiltration of corrosive salts and sediments clog sewage outlets and drainpipes, which then burst within the building walls.

BASILICA

FOUNDED in 829 AD, in honour of the city's patron saint, St Mark's Basilica is the city's most famous building, and one of its most vulnerable. Its crypt now lies some 20cm below sea level, nearly 170cm lower than when it was built. Water invades the atrium between 150 and 180 days a year (nearly every day in winter), when the tide touches 67cm above reference zero. As water level rises, it starts to seep up through the foundations, while outside in St Mark's Square, it comes up out of the storm drains to flood the pavements. In winter, the vestibule is commonly ankle-deep in water. The floor of the Basilica has been raised many times in the past, and the square has been repaved. But today this is no longer an option: further raising would destroy the architectural quality of this magnificent church. Other remedies must be found (Chapter 4 explores works underway in greater detail).

CHAPTER 3 SUMMARY

■ Flooding is caused primarily by storm surges – winds that drive water into the lagoon

■ Many man-made changes have inadvertently reduced the lagoon's resistance to incoming waters, against a background of sea level rise – hence higher tides

■ It is not a new phenomenon for Venice, but the damaging effects of flooding create an insatiable demand for urban maintenance and are gradually driving the population away

■ The idea that Venice is sinking comes from the extraction of groundwater in the mid-20th century, which caused major land subsidence. This has now been stopped and Venice has returned to subsiding naturally and gradually

■ Barriers at the inlets will protect Venice from extreme water levels; other measures are being taken to protect the city from medium-high tides and 'chronic' flooding

Remedies

"Our goals are to restore and re-equilibrate the lagoon, and eliminate the causes of environmental degradation"

Maria Giovanna Piva, President, Venice Water Authority

THE complex problems of Venice and its lagoon must be understood as a whole, since the health of one depends critically on the health of the other. An integrated programme of defences has been outlined to protect Venice from the *acque alte*, and to reverse the degradation of the lagoon environment. The urgent and constant need for interventions to restore and preserve the fabric of both city and lagoon, and to protect them from floodwaters, go hand in hand. There are enormous technical challenges, which involve establishing priorities, recognising the different timescales for action and managing uncertainty. Much of the work involves tested engineering approaches but some of it is experimental in nature.

THIS CHAPTER EXPLAINS
- **'LOCAL' MEASURES TO PROTECT VENICE AND THE OTHER ISLANDS**
- **THE RESTORATION OF VENICE'S BUILDINGS AND INFRASTRUCTURE**
- **BUILDING NEW SALTMARSHES; THE RESTORATION AND CONSERVATION OF REMAINING ONES**
- **OTHER 'DIFFUSE' MEASURES IN THE LAGOON**
- **THE NEED FOR FLOOD CONTROL OVER AND ABOVE THESE MEASURES**

left Building maintenance is a constant necessity from the foundations upwards

Defending the city

Protecting Venice from flooding requires a high degree of planning and coordination. The current programme of works is also a great opportunity to renew the city's infrastructure

AN integrated system of defences has been outlined to protect Venice from the *acqua alte*. These are classified into:

- 'local' (city and other inhabited islands)
- 'diffuse' (lagoon-wide)
- 'coastal' and large-scale engineering (the barrier scheme and shoreline reinforcement).

The *Comitatone* determines safeguarding policy, while the Consorzio Venezia Nuova is charged by the Venice Water Authority with planning, designing and implementing them – as far as the lagoon is concerned. Insula SpA does the same for the urban areas of Venice and the islands. Both work with regional and local authorities and other key bodies.

The proposed scope of works is vast: embankments and walls are being raised against the rising waters; canals require dredging; kilometres of buildings and canal embankments need to be restored from the foundations up; the entire sewage system and underground drains need to be repaired and modernised. Out in the lagoon, the steep decline in physical and ecological status has to be redressed, in order to restore the natural resilience of the system, and new flood protection measures must be put in place. At the coast, the sea walls need strengthening, while flood barriers and associated works are to be built. It is an extraordinary, ambitious programme, calling for more integrated analysis of the past 30 years' acquired scientific knowledge, sharing of data and careful monitoring of the effects and impacts of interventions – during and after implementation.

DEBATING THE PRIORITIES

There is considerable debate in Venice about which of these safeguarding measures should take priority, and indeed whether they are all necessary to protect the city. Some scientists and engineers feel that the mobile barriers (discussed in Chapter 5) need to be built immediately to eliminate the risk of another devastating flood like 1966. They

argue that tackling environmental degradation of the lagoon can be dealt with separately over a longer timescale. Others, who oppose the barriers, would prefer to see the local and diffuse measures prioritised to combat the chronic flooding problem and deal with the environmental degradation.

above **Ad hoc** measures to keep out the floodwaters can be seen everywhere in Venice

LOCAL DEFENCES

Local defences are designed to protect the built-up areas from flooding. The first of these measures involves raising pavements, bridges, embankments and walkways as well as subsidising anti-

left Archaeologists have uncovered evidence throughout the city of repeated raising of floor levels to beat the rising waters. Here a column's base is now many centimetres below the current floor level

flooding measures for the ground floor of many buildings, especially in the lowest-lying areas of the city and other islands. The stated aim is to defend the city against medium-high flooding tides (up to 110cm above the reference level) and, where possible as high as 120cm.

Historically, Venetians have responded to rising water levels by knocking down old buildings and building new ones, or by simply filling in canals, and raising the floors and pavements. There has been much discussion about whether it is possible just to continue this habit to keep pace with the waters. But, in the opinion of many experts, the limit has been reached if the architectural glories of Venice are to be preserved, as raising pavements any further will destroy the elegant proportions of the buildings and eliminate features such as the bases of columns. Doing

so would create a different Venice, and it is not an option that most people are prepared to accept for this World Heritage Site. With local defences only of the order of 110cm, the barriers will be vital in defending Venice's iconic buildings and residents from extreme floods.

Another category of local defence works is known as *insulae* (islets), which go beyond simply building up ground levels and include the impermeabilisation of an entire islet area within the city, via complex interventions on the drainage system, use of one-way valves and special materials to block the seepage of underground waters.

More ambitious schemes to raise whole buildings or entire areas of land using innovative technologies are still technically uncertain, but they remain a long-term option (See Chapter 6).

ARCHITECTURAL SURVEYS

IN 1999, the *Comitatone* requested clarification on the real potential to raise ground levels throughout Venice, "tending towards 120cm". Studies were commissioned by the *Comune* (Insula and COSES) and by MAV-CVN (IUAV-ISP, a consultancy group within Venice's Architecture University). Each study group adopted a different approach and set of parameters, and their conclusions were different. IUAV did an intricately detailed survey of three areas of the city and highlighted the difficulties of generalised ground raising against a background of highly variable and intricate architectonic and aesthetic building features.

Insula, on the other hand, concluded that it would be realistic to aim for local defences of +120cm relative to water level, although the details of each intervention must be evaluated on a case by case basis. Insula also applies a less restrictive consideration of architectural features which allows, for example, steps to be covered by higher paving and architraves to be knocked out to allow the height of a doorway to be increased to compensate for a higher threshold. Both analyses represent an extraordinary record of the city today, and a powerful resource for the future.

To date, Insula has completed more than 15 percent of the 100,000m² earmarked for raising and it has also taken a further 15,000m² of walkways which were already at +110cm to +120cm or above. On average it has added 8cm to relative height, which

corresponds to a 50 percent reduction in flooding frequency. A further 30,000m² of embankments, including the islet of St Mark's has been assigned to MAV-CVN.

above **Analyses of the proportions of doorways, and how they might be modified if pavements were raised. Insula considered 13,000 units with less restrictive parameters and IUAV surveyed 9,300 units with great attention to retaining distinctive architectural and historic features**

Repairing and renewing the city

Work to rescue Venice's crumbling fabric was given new impetus with the Special Laws, which led to the development of a large-scale programme of restoration, renewal and flood protection

above Replacing crumbling brickwork

right Laying new electricity and other service networks across the city

MAINTAINING Venice has been a constant challenge ever since the city was built — water and sea air are naturally corrosive, while modern pollution also attacks marble and plaster. Venetians have worked hard to preserve their city – although urban maintenance took a back seat in the post-World War Two period, efforts were intensified following the shock of 1966. Despite the past 30 years' work, there is still much to be done to restore and maintain the city's urban fabric.

In the 1990s, Venice Town Council linked up with key service industries – water, gas, electricity and telecommunications – to form a company called Insula, to develop and run a massive integrated project to protect and restore the city. Raising ground levels for local flood protection was synergised with renewing the city's ageing root-and-branch infrastructure, and dredging the canals.

Periodic dredging operations have been performed since ancient times to permit navigation and to maintain acceptable hygienic conditions. From the early 1960s to late 1990s, however, the dredging was interrupted, except for some minor interventions, and a layer of sediment up to 1m thick accumulated in the network. Canal dredging to improve water circulation and navigability is now combined with repairing building foundations, increasing the number of septic tanks and the renewal of the culvert system. Disruption to the city's life is enormous – pipelines are being laid, bridges and walkways are being rebuilt, canals are drained while workmen replace and consolidate the brickwork and inject it with innovative chemicals.

Once canals are dredged, the foundations are accessible and repair work is more easily done. An extensive programme of work on dredging and repairing the canals is underway, and urgently needed modernisation of the sewage network, to include a collection and treatment system for

left Repairing a sewage outlet

below The canals are dredged to get at the foundations and underwater sewage drains

CURING SALT DAMAGE

Salt-damaged brickwork

Workmen injecting brickwork with consolidating and stablising materials

SALTWATER and sea air attack both above and below the water line. In the past, Venetian builders created a damp-course at the base of their buildings, using non-porous Istrian stone. The underwater depth of the stone facing was determined by the then high tides, with a built-in safety margin. Now that margin has been lost. Saltwater penetrates and rises in the brickwork. Once the water evaporates, the salt left in the wall crystallises, expanding and so increasing pressure on the bricks, which crumble along with plasterwork. Extensive research has resulted in the development of innovative chemical and physical treatments to protect the brickwork and block salt invasion. Some verge on pure alchemy, others are simpler – for example the use of outer plasterwork as a 'sacrificial layer' in which the salts concentrate and crystallise, rather than staying within the brickwork.

the historical centre, is being discussed. Venice has no sewage treatment: traditionally the city discharges its waste untreated into the lagoon, relying on the tides to flush the canals clean. But today many wastewater outlets, constructed centuries ago, are covered by mud in undredged canals or clogged by fine sediments, stirred up by boat traffic and the tides. This weakens buildings as the old pipes disintegrate and release their contents within the walls. Fluctuations in water pressure can also damage foundations and canal banks. On the mainland around the lagoon, new treatment plants and modern sewers are already being put in place to stop the influx of nutrient-loaded sewage into the lagoon.

Rescuing St Mark's Square

ST MARK's Square, Venice's iconic landmark and symbol of its past glory, is the target of a special programme of flood protection and repair. It is an example of the *insulae* approach to improving defences against medium-high tides. Work is proceeding in small sections, so that the Square remains accessible to the thousands of daily visitors.

Phase One, started in 2003, consists of repairing and raising 150m of quayside walls to +110cm to stop waves and floodwaters from flowing over the banks. The pavement within the Square cannot be raised any further without ruining the architectural proportions of its unique buildings.

Phase Two consists of reorganising and modernising the existing network of pipes and drains underneath the Square, and fitting one-way valves to stop the backflow of water up to the surface. Underground conduits will be completely restored where they have collapsed. A waterproof layer may be added to prevent seepage through the subsoil, but this is subject to further research. A new rainwater collection system linked to a pumping station will actively pump water back out to the lagoon at high tides.

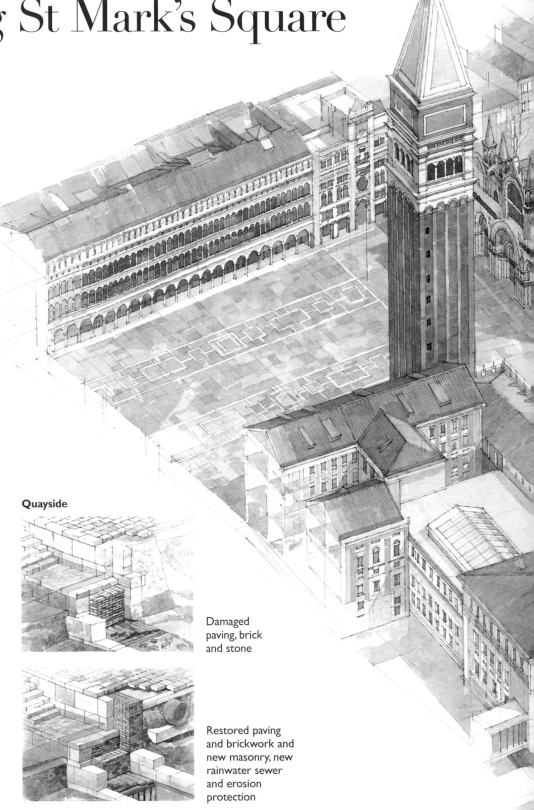

Quayside

Damaged paving, brick and stone

Restored paving and brickwork and new masonry, new rainwater sewer and erosion protection

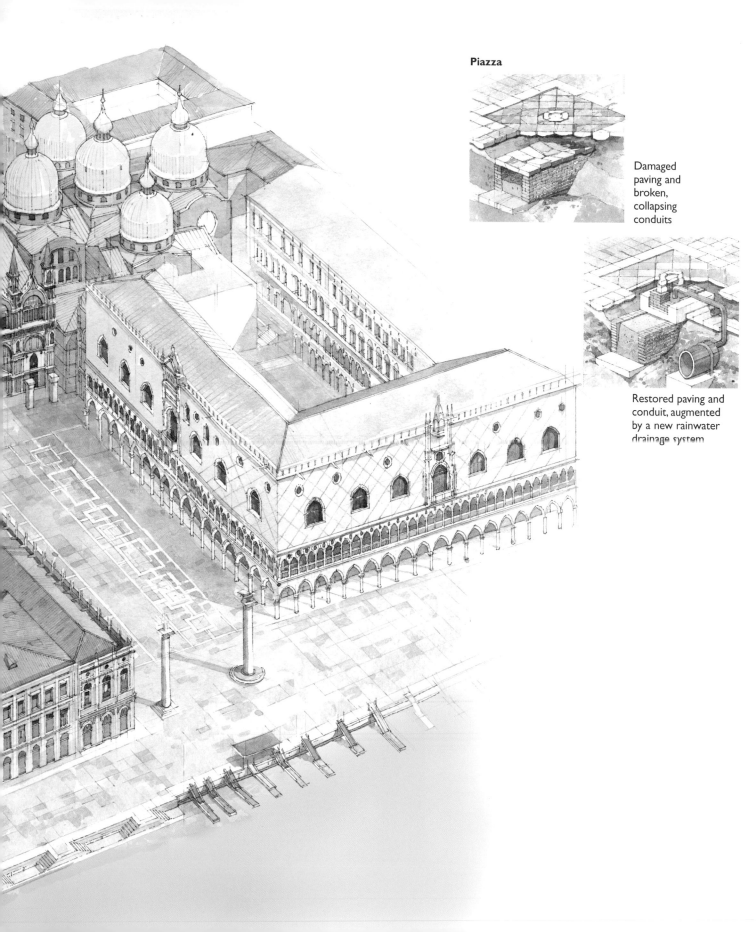

Piazza

Damaged paving and broken, collapsing conduits

Restored paving and conduit, augmented by a new rainwater drainage system

Protecting the lagoon

A second layer of 'diffuse' measures covers flood defences and environmental restoration across the lagoon. They help to protect Venice, mitigate the tides and revive the lagoon's degrading habitats

EROSION, high waters and wave motion are gradually transforming the lagoon into a marine bay. It is losing its physical characteristics and habitats. The 'diffuse' measures proposed across the lagoon seek to reduce the height and impact of high tides and, in addition, to restore the lagoon morphology. This second line of defence aims to restore the natural defences that have made possible Venice's long history, working with nature rather than against it. The lagoon's physical and biological structures used to help dampen the effect of *acqua alta* but over the last century so much has been damaged and eroded that the seas flow more freely into the lagoon and easily penetrate the canals of Venice. The measures will have an impact on the water levels but cannot solve the flooding problem, although they may reduce the frequency of flooding. They will, however, play an important role in reversing the degradation of the lagoon.

The 'diffuse' measures include works on salt-marshes, further investigation and pilot studies to quantify the potential of opening areas to tidal exchange, filling-in dredged navigation canals and modifying inlets. Additional measures aimed at protecting the lagoon rather than influencing flooding include improving the ecological quality in some areas; controlling the release of pollution from sediments and especially from the dump sites where industrial wastes were previously disposed; planting seagrass to help consolidate muds; improving water quality; protecting and reintroducing saltmarshes. There is no one solution to protecting Venice and its lagoon as a combination of approaches is needed. Also, these proposed measures are not all feasible, but do form part of the debate of which remedies should be undertaken as a priority and over what timescale.

Revitalising the saltmarshes

In the debate on the future of Venice lagoon there seems to be a common aim: guaranteeing the conservation of this unique system. Experimental restoration work is taking place in 20 locations

AS described in Chapter 2, the lagoon's saltmarshes and mudflats are still disappearing at an alarming rate as a result of reclamation, erosion, pollution, and natural and human-induced subsidence. Reconstructing lost saltmarshes and protecting and restoring the existing ones is a major technical challenge. Their natural forms are the result of evolving conditions, an intricate balance of interacting processes – the energy of the waves and the ability of the structures to dampen them, the amount and flow of sediment in the system, the biological mix of species. Any attempts to reconstruct or restore these habitats must be based on an understanding of how the natural system works. It requires innovative thinking – and many of the techniques being tried out are highly experimental. It is, anyway, impossible to return to the lagoon as it once was: it is deeper and more open to waves than before, and the dynamics of flow are now quite different.

RESTORING OLD MARSHES

To restore the existing marshes to health, it is vital to protect their eroding edges and give them time to build up again. A range of experimental techniques are being trialled, using wooden piles, gabions (wire cages containing rock or other material), artificial sandbars and beaches. Other experiments include measures to encourage sedimentation and new growth: sediment fences, spraying the marsh with nutrients and depositing new organic matter at the edges, and transplanting vegetation. Water quality can be improved by re-opening tidal creeks. Polluted mudflats are being capped to prevent their sediments being stirred up into the water column.

CREATING NEW MARSHES

In some areas new marshland is being created using dredged material, taken from navigation channels in the lagoon and near the inlets. By the end of 2000, 15 percent of total 'marshes' in the

lagoon was now of this type. These are highly managed systems, with solid edges to prevent the unstable sediment from being washed away again by the waves. There is already concern that their steep edges mean they are not linked with mudflats that grade into natural saltmarshes. They are distinctly different in character from the other marsh areas, their structure a 'cruder' version, with less physical and biological complexity. The question is whether they develop into stable, complex communities over time which are comparable to the Venetian saltmarshes.

These are long-term experiments, and the results will not be apparent for decades. They

require continual monitoring and testing by scientists to follow their progress. Lessons have already been learnt. For example, there is now a move away from using hard structures to using degradable materials and engineering solutions that are closer to the natural structures and processes.

The illustration over the page shows some of the key restoration projects. As a summary diagram, it does not reflect any one place – these measures are being tried out in many different locations.

above Natural saltmarsh

far, left Fishermen set out their nets in the open lagoon and the marshes. Fish such as grey mullet and flounder are caught when they become trapped by the changing tides

HYPOTHETICAL MEASURES

Re-opening areas of the lagoon

IN the last century, the area of lagoon open to the tides shrank by about 20 percent. The fish farms, situated at the lagoon periphery to the north and south, are now separated from the lagoon by embankments, isolated from the ebb and flow of the tide. Other areas of the lagoon were made into artificial islands, mainly in the 1960s, in anticipation of an industrial expansion that never happened. These are to be opened up by digging tidal creeks through them, and planting them with vegetation to encourage the return of saltmarshes, as part of the Fusina water treatment system.

Opening these areas would allow the tidal waters to be dissipated over a larger area. But studies have shown that it would only slightly reduce water levels. Also, the farms are protected by powerful interest groups, including recreational hunting.

Navigation channels

THERE has also been much debate about filling in the major navigation channels (notably the *Canale dei Petroli* that links the Adriatic to Marghera, via the Malamocco Inlet) which are widely considered to be a major cause of erosion processes in the central lagoon. At the moment, inter-tidal sediments are resuspended by the influence of the currents in the deep channels, which is where they re-deposit, only to be lost from the system as they are carried out to sea by the strong currents.

Filling in the channels would reduce the amount of sediment lost from the inter-tidal areas and impede the dissipation of water entering the lagoon at high tide. But these channels carry important traffic to and from the Port of Marghera, and there is equal pressure to keep them open and dredged – studies have shown that filling-in the *Canali dei Petroli* would have little influence on the water level and flooding of Venice.

Container ship off-loading

Experiments in restoring saltmarsh

THIS summary illustration shows a range of restoration techniques being tried out around the lagoon. Applications need to be monitored for their effectiveness in developing stable, healthy saltmarsh communities in order to guide future projects. Techniques are being shared around the world, especially with the North Sea (Germany, Holland and UK) and Chesapeake Bay (USA).

Protecting the edges

The areas where restoration is being trialled are afflicted by **eroding edges**, where the deepening lagoon and greater expanse of open waters produce higher and more powerful waves, which batter the marsh banks. Invasive seaweeds have also killed off vegetation that helped protect the margins via their root systems. **Eroding shallows** are another problem, as the disappearing mudflats used to dampen the force of the waves and supply sediments and ecological relationships to nourish the marshes.

A Hard 'gabions' (wired crates of stones) have been tried in more exposed areas, but can cause local wave action and thus increase local erosion. They also interrupt the natural process of exchange between the marsh and adjacent mudflats.

B Biodegradable gabions (sacks of sediment and/or bark) planted with pioneer vegetation are designed to dampen wave forces while they degrade gradually, allowing natural protective saltmarsh regrowth to take their place.

C Simple woven brushwood fences help to protect edges and increase sediment trapping by 1–3cm a year. They require constant maintenance.

D Wooden pilings are considered an alternative to hard gabions.

Restoring the vegetation

E Replanting helps to stabilise the lagoon bed – by trapping sediment in the shallows and triggering the process of rebuilding the marsh.

Feeding the marshes

F Boats can be used to replenish marshes with a fine spray of sediment.

Dredging the tidal creeks

G Excavating is done to improve flow through the marshes, to facilitate water and nutrient exchange.

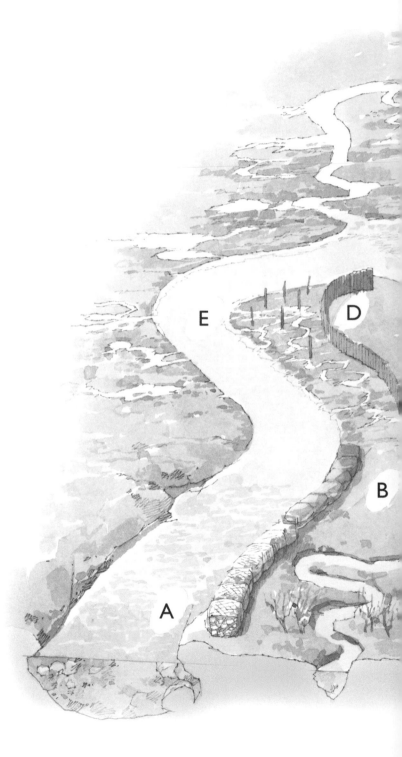

Measures at the coast

The proposed programme of protective measures continues at the coast, to restore existing defences and alter the inlets to increase their dampening effect on the incoming tides

THE coast is the lagoon's first natural defence against the sea — waves lose their energy as they break on beaches that form a barrier across the lagoon. But erosion has progressively reduced the beaches to the point where some have disappeared altogether, as has the dune belt behind them. These natural structures reduce wind action and trap windblown sand, and are also important wildlife landscapes. There are numerous small towns and villages along the Venice coast, making protection here even more vital.

Coastal defences include creating new beaches and widening existing ones. The ancient sea walls and jetties require maintenance, and new groynes are being built out from the beaches to dampen the effect of storm waves. Sand dunes also need to be conserved and preserved, with the added benefit of bringing wildlife back to the bay.

TIDAL INLETS

There has been much debate about how to alter the three tidal inlets, at Chioggia, Malamocco and the Lido, to restrict the amount of water entering the lagoon. Some scientists have supported proposals to narrow the inlets and where possible to reduce the depth and roughen the seabed there. This would create higher resistance against water flowing into the lagoon during a strong surge, resulting in a lower volume of incoming water over a given period of time and a lower peak water level. The effect of inlet modifications would be less for normal tides and almost negligible in a 1966 type event.

Other scientists argue whether this approach is useful as part of the overall safeguarding measures due to long-term impacts on the water flows and sediment processes. The *Comitatone* agreed in 2003 that some changes to the inlets will occur as part of the so-called 'complementary measures' discussed in the next chapter.

A HUGE CHALLENGE

The task is a vastly challenging one, as each change

made to the lagoon will ripple through the system. Many measures are experimental and what works 'on paper' may not work in practice in such a complex, multi-dimensional system. Detailed research and constant monitoring is needed to track the changes and measure their impact.

At the same time, scientists are still working to establish a sufficiently detailed baseline of information about the lagoon's behaviour today. Without this, they cannot accurately judge the true impact of the changes.

IS IT ENOUGH?

Could the local and diffuse measures described in this chapter be sufficient to protect Venice from the floods?

While a few people argue that they will, the majority of scientific opinion is that, although they help limit the number of floods, they are not sufficient – they cannot protect the lagoon from extreme

left The inlets are demarcated by rock and concrete jetties

right A beach was artificially built at Pellestrina to give greater defence against the sea

below right Pipelines are used to deliver a sand-and-water mixture for beach replenishment

events, like the storm of 1966. Nor can they offer protection from inevitable long-term global sea level rise. Studies aided by computer simulations suggest that neither the lagoon-wide restoration efforts nor the changes to the lagoon inlets (presently under construction) will be able to reduce high water levels in Venice by more than a few centimetres.

Measures in the city and across the lagoon would have to be far more drastic, and unacceptable in view of the world importance of its architectural heritage. And most people are certain that it is only a matter of time before another major disaster happens.

The consensus is that the only feasible way to stop the rising waters from overwhelming the city again is to provide a physical separation from the sea.

CHAPTER 4 SUMMARY

- **All projects to restore the lagoon and protect the city must work in a symbiotic way**
- **There are two kinds of measures: 'local' measures within the city and the islands, and 'diffuse' measures across the entire lagoon environment**
- **Local measures offer vital opportunities to improve the city's infrastructure and restore its buildings**
- **Some remedial measures are highly experimental and their effects will only be learned in the long-term**
- **Restoring the saltmarshes is key to the future health of the environment, its unique wildlife and the resilience of the entire lagoon system**
- **Most scientific authorities are in agreement that the most viable way to protect Venice from extreme flood events is to create a barrier**

Barrier

"Building barriers and flood defences against changing sea levels is an on-going process. Barriers buy you time"

Sarah Lavery, Thames Flood Defence Strategy 2100

THE mobile flood barriers to be built at the three inlets to the Venice lagoon have a single purpose, to protect the city and lagoon islands from major storm surge flooding. They are being developed against a background of intense scientific and political debate about whether they should take priority over other measures, how viable they are, and what their likely impact will be on the life of the lagoon. The barrier design is radical and poses enormous engineering challenges. Research will continue as the gates are built and tested, as will research into their effects on the natural system of the lagoon. Lessons can be learned from other flood barriers across the world, but each barrier system and location is unique.

THIS CHAPTER EXPLORES
- **THE CONSTRUCTION OF THE BARRIERS**
- **THEIR MECHANISM AND OPERATION**
- **THE COMPLEMENTARY MEASURES AT THE COAST AND INLETS**
- **OTHER BARRIER SYSTEMS**
- **THEIR PREDICTED IMPACT ON THE LAGOON SYSTEM**
- **DEBATES ABOUT THEIR ENVIRONMENTAL IMPACT**

left Malamocco inlet, where construction of the breakwaters associated with the barriers officially began in spring 2003

The mobile barriers

THE barriers will consist of 78 hinged steel floodgates, stretching across the three inlets of the lagoon. The wider Lido inlet barrier (41 gates) will be in two parts, separated by an artificial island. The Malamocco (19 gates) and Chioggia (18 gates) inlets will each incorporate a navigational lock for shipping.

The huge gates are 20m wide, up to 5m thick and between 18 and 28m in length (depth). They will be anchored at one hinged end to concrete foundations sunk into the lagoon bed. They will lie unseen on the bed, filled with water, until a storm warning comes, when they will be pumped full of air and rise to form a wall against the sea. They will remain up for the duration of the surge tide, and will then refill with water and sink down again.

The construction of the barriers is a major engineering feat, not just to design a complex mechanism that will continue to work reliably for decades, but also to manoeuvre the huge panels into place without error or technical hitches. The materials used must withstand the harsh conditions of the saline waters.

The proposed mobile barrier. Design modifications, even during construction, are a common and expected part of sophisticated engineering work, and the final design is likely to vary slightly from this specification, first published in 1996

Location of barriers at inlets

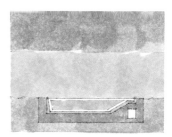

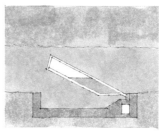

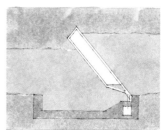

How the individual panels
will rise and fall

How will the barriers work?

THE idea of a system of gates between the lagoon and the Adriatic Sea is not new – a scheme was even proposed in the 17th century. But the need for protection came sharply into focus after the 1966 flood and in 1971, a 'competition of ideas' was held. The concept that won was for an underwater mobile barrier.

It took more than a decade for an institutional framework to design, assess and start to build the barrier to be put in place. The 1984 'Special Law for Venice' led to the creation of the *Consorzio Venezia Nuova* (CVN), a consortium of large Italian engineering and construction companies, which was charged with implementing the solution to the 'water problem' (for details, see Appendix).

Twenty years of planning, technical development and testing followed, amid intense scientific and political debate. The final go-ahead to plan and build the mobile barriers was given only in 2001, and work began in earnest in 2003 with the construction of an outer breakwater at Malamocco.

The first experimental prototype, called MOSE (*MOdulo Sperimentale Elettromeccanico*) was built and tested between 1988 and 1992. It has given its name to the larger scheme – the complete set of mobile barriers and 'complementary measures' at the inlets.

MOSE is being developed under a set of constraints imposed by the *Comitatone* (see Appendix), deriving from environmental assessments in the late 1990s and pressure from the Port of Venice and industry concerned by the possibility of disruption to their transport links. The barriers must have little or no negative impact on navigation, nor on lagoon flushing and water quality, ecology and habitats, or on the 'aesthetics' of the lagoon. The latter is a key reason for the underwater design solution.

The project is being developed through detailed engineering studies, physical modelling and computer simulations to explore how the barriers will work *in situ*, how they should be managed, and to assess their impact on water flow in the lagoon. This work must be integrated with, and be informed by, scientific studies on the

lagoon's physical and biological functioning, as well as research to improve weather forecasting. It is remarkable that an enormous amount of data collected by numerous institutions (but mainly the Venice Water Authority) has not been circulated or made readily accessible to the scientific community nor the public at large. This has obstructed the comparison of research results and findings from different modelling approaches and the development of a non-ideological, constructive, science-based debate on certain key issues.

FLOOD ALERT

When water levels of more than 110cm above the reference level are predicted, the mobile barriers

"This type of project requires a high percentage of unusual work and advanced technology."

Alberto Scotti, barrier design engineer

above Engineering interventions at the inlets were also considered 400 years ago

below A physical model of the lagoon is used to investigate the system and the barrier impacts

will be put into operation and stay in place for the length of the tide surge. On flood alert days, the lagoon will be cut off from the Adriatic Sea for an average of five hours at a stretch (longer if there are continuing surges in the following days as was the case in 1966). The imperative is to close the gates at the right time, for the shortest time possible.

Deciding when to close the barrier is a complex matter, as not only the tide affects water levels in the lagoon but also the weather (the atmospheric pressure, winds and rain). Some water will still flow into the lagoon even when the gates are up, and heavy rains will also bring water into the system. In this case, the gates must be activated when sea level is still relatively low to allow a margin for internal water level rise. Winds can drive high waves across the lagoon, meaning that water levels may be higher in some areas than others.

A reliable model is therefore essential to predict and guide barrier closure decisions. Present simulations can result in 50 percent 'false alarms' and more research will be needed to improve accuracy.

PROPOSED MODIFICATIONS AT THE THREE INLETS

These diagrams show the position of the barriers and navigation locks at the three inlets. The MOSE scheme is required to reduce the inflow of water from the Adriatic Sea by between 15 and 28 percent at the inlets. To achieve this, the complementary measures – moon-shaped (*lunate*) breakwaters will also be built beyond each inlet. The Malamocco outer sea wall is about 1,300m long and 3–4m above mean sea level; at Chioggia it will be about half as long and 2.5m high.

The Malamocco breakwaters will be complete by the end of 2004 and work is beginning at Chioggia. The Lido breakwaters are still at the design and approval stage. At the same time, the channels through the inlets will be reduced in depth and their profile will be changed. Some engineers believe that these changes will help increase resistance to the flow of water in and out of the lagoon, and so further reduce the volume of water entering from the Adriatic Sea. Other hydraulic engineers expect the benefits to be slight, and the key issue is cost effectiveness. There is also concern about the long-term ecological and water quality impacts of these changes.

Lido inlet
800m wide, 12m deep, two barrier systems and a central island

1 Boat haven
2,4 barriers
3 artificial island
5,6 existing jetties
7 breakwaters

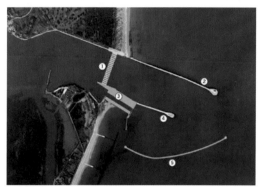

Malamocco inlet
400m wide, 16m deep

1 Barriers
2,4 existing jetties
3 navigational lock
5 breakwaters

Chioggia inlet
380m wide, 10m deep

1 Boat haven
2,4 existing jetties
3 barriers
5 breakwaters

WHAT WILL HAPPEN TO TANKER TRAFFIC?

The navigational locks allow shipping to continue undisturbed by gate closure, except where stormy weather prevents them coming into port anyway and large ships may find it more difficult to steer past the breakwaters and through the new locks. But computer simulations have shown that it will be possible. In the future oil tankers may not have to come into the lagoon – there have been proposals to locate an oil terminal out in the Adriatic Sea and link it to the port via a pipeline, removing the threat of an ecologically disastrous accidental spill in the lagoon.

Learning from other schemes

The successes and failures of other schemes have valuable lessons to teach Venice. In turn, Venice's experience will inform other schemes

FLOOD barriers protect many vulnerable cities in Europe. The three examples given here – from London, Rotterdam and St Petersburg – highlight some of the experience that Venice can benefit from, although the context and tidal systems differ markedly and every situation bears a certain uniqueness. Each project was set up to protect major cities and other areas of dense population from sea storm surges, and each scheme chose a completely different barrier system.

In two cases, there was serious conflict between environmental impact and the desire to protect the urban population. In all cases, major social and economic issues were at stake. A balance always has to be struck, a trade-off between protection, trade and environment. Everyone also has to recognise the scale and uncertainty of these grand schemes, where whole ecosystems are dramatically changed – indeed, in the Dutch case, altered beyond recognition from salt to freshwater. Research cannot stop when the barrier is built; the consequences of our actions need to be understood, so that we learn for the future.

In summary, the lessons to be learned are that:

- Flood disasters are commonly a trigger for action
- There needs to be a general public and political will to drive the project forward – often it is only crisis that focuses people's attention
- Major engineering schemes are expensive, experimental and difficult
- Funding needs to be secure – half-completed schemes pose environmental hazards in themselves

left Other European flood barriers
1 Thames Barrier
2 Rotterdam Waterway
3 St Petersburg

- Barriers cost a lot of money to operate and maintain
- High-tech barriers need skilled people to run them, not just to build them
- Design modifications are normal as engineers learn more about how a barrier will perform once installed, and what impact it will have
- There will inevitably be a trade-off between safety and environment, the skill is to find a way of working together, rather than in opposition
- In a changing world, barriers are never a final solution – cities have to keep planning for the following 100 years

The Science of Saving Venice

The Thames Barrier system is now also used for management of river flooding – an unforeseen application at the time of construction

THAMES BARRIER, LONDON, ENGLAND

The River Thames runs through a densely populated and commercially sensitive floodplain, including the City of London, and flooding has the potential to cause economic and social disaster. Surge tides from the northern Atlantic are forced down into the North Sea, driving extra water up the river and occasionally flood defences are overtopped. Plans to protect the city were triggered by a disastrous flood in 1953, which killed 317 people in England and Scotland.

The chosen barrier scheme consists of a series of rotating (normally submerged) gates suspended between a series of towers. Construction started in 1974 and the barrier opened ten years later. London now has one of the best tidal defences in the world, designed to cope with storm surges and estimated sea level rise well beyond 2030. But climate change is predicted to bring higher seas and uncertain changes in storminess, and planning has already begun for the 'next generation' of protection up to 2100.

Learning from other schemes continued

The Maeslant defences are controlled by a sophisticated computer system

The Science of Saving Venice

ROTTERDAM AND THE EASTERN SCHELDT, HOLLAND

Holland is particularly vulnerable to flooding, some 60 percent of the land being below sea level, so flood defences are critical. Just as in England, the decision to embark on large-scale storm and flood defence systems to protect the area around Rotterdam was taken following the disastrous 1953 flood, in which 1,835 people died in Holland. This led to the 'Delta Plan' to reinforce the dikes and to close many of the existing estuaries with dams.

Only two of the estuaries were left open: the Western Scheldt estuary, which leads to the Port of Antwerp, and the New Waterway, which is the entrance for the Port of Rotterdam. For the New Waterway, a barrier project was started in 1991, and just six years later the Maeslant barrier was completed. This barrier is almost as wide as the Eiffel Tower is high. Giant swivel gates swing across the estuary to form an impenetrable wall protecting one million people.

During its construction, the engineering design had to be revised when the team realised that 'seiches', long waves with a period of 15–90 minutes and with amplitudes of over 2m, could create a negative head over the gates, which were initially not designed to resist them. Seiches are also a feature of the Adriatic, and the Venice gates will have to cope with them.

The Delta Plan brought about considerable changes in the tide and the ecosystems of the area. In some areas the decision was made to change the ecosystem completely from a saline to freshwater environment. At the Eastern Scheldt, there were furious protests culminating in a sea battle pitting environmentalists and fishermen against visiting politicians. A much more expensive system resulted, incorporating a storm surge barrier.

The Maeslant barrier has now been operating for seven years, and has yet to be closed for a surge emergency. But it is tested yearly, just before the beginning of the each storm season, to check it still works. Dutch experts stress the importance but difficulty of maintaining highly skilled teams, who are used to the demands and vulnerability of such a high-tech system.

Initial attempts to build the barrier suffered from poor public consultation

ST PETERSBURG, RUSSIA

Like Venice, St Petersburg is a major port, and one of Russia's greatest architectural treasures. The city is in a low-lying coastal plain where the Neva river discharges into the Gulf of Finland. Much of the old city lies little more than 2.5m above the sea, and floods have been part of its history – the greatest in 1824 when waters rose nearly 4m above normal levels.

Plans to build a 24km flood barrier across the Neva Bay began in the 1960s. Construction started in 1980, but work halted in 1990. There was also huge local and international opposition to the scheme, based on the fear that the ecology of the bay and the Gulf of Finland would be seriously damaged by the system. The shallow waters are internationally important breeding grounds for fish and feeding grounds for migrant birds.

Extensive studies have since shown that the bay is already suffering serious environmental problems due to pollution from the city and industry. Large-scale field experiments have since demonstrated that the gates could be used to change the water flow in the bay, to help rather than damage the ecology. This has also been discussed in Venice.

In 2002, the scheme restarted when a consortium of international banks agreed to support completion of the flood defence project. A public consultation exercise on the environmental aspects of the scheme resulted in both an improved feasibility study and increased local support for the barrier. The barrier should be completed in five to seven years.

Debates about the barrier

The key debates around the barrier scheme concern the possible impacts on the water quality and ecology of the lagoon and the cost-effectiveness of the scheme. Some uncertainty prevails

ONE of the most serious political challenges to the MOSE scheme has come from environmentalists and their scientific supporters, who believe that the gates will have a very damaging effect on the lagoon. MOSE is now underway, but the questions still come.

HOW OFTEN WILL THEY HAVE TO CLOSE?
At present, the barrier designers predict that the gates will operate an average of five times a year. As storms tend to occur in winter, most of these closures will be concentrated in a few months. The gates may stay closed for several tidal cycles when the storm surge is followed by echoing seiches.

ARE THESE ESTIMATES RELIABLE?
The opponents argue that predictions of how often the gates would operate are unrealistically optimistic, as global environmental change is expected to bring higher sea levels and may increase future storminess.

HOW WILL THE LAGOON BE AFFECTED?
The basic argument of opponents is that, when the floodgates are closed, the lagoon will become a closed basin, cut off from the beneficial effects of tidal flushing. The lagoon will also be cut off from sediment exchange during a storm. Although we still do not know enough about the current role of storms in the net sediment transport for this system, it is possible that the barrier closure could have a positive effect on reducing sediment export from the lagoon.

WHAT ABOUT THE WATER QUALITY?
Without daily tidal flushing, ecologists fear that the water in the lagoon will stagnate and bottle up pollutants and nutrients. Fish and other species could suffer, and the algal blooms that plagued the lagoon in the 1980s could return. The counter-argument is that flushing will occur very quickly once the gates re-open – recent studies

have shown that much more water is exchanged between sea and lagoon on each tide than was previously thought.

COULD THE GATES ACTUALLY HELP WATER QUALITY?
Some scientists and engineers have even proposed that the barriers could be operated, when not needed for flood control, to *stimulate* flushing in the lagoon. By opening and closing the gates in sequence, for example, favourable water circulation patterns could be generated within the lagoon to aid flushing. However, it is also recommended that a correct solution is, after all, to tackle the problems of water quality and pollution at source, such as providing a sewage treatment system for Venice, and not to look to the sea to solve them.

"Glory, if it succeeds, and grave responsibility, if it fails, for everyone"

Local newspaper

The Science of Saving Venice

WHAT EFFECT WILL THE BREAKWATERS HAVE?

Some scientists claim that it is not the barriers that present the problem, but the *lunate* breakwaters – their effect on dissipating the inflow of the sea may be fairly marginal plus they could create additional sediment budget complications.

AREN'T THERE ANY OTHER SOLUTIONS?

Many other barrier concepts have been proposed over the years, with varying degrees of professional credibility and feasibility testing. They include placing a number of smaller flood barriers at the entrance to the main island canals; installing inflatable rubber tubes at the inlets; and parking old tankers filled with water at the bottom of inlets instead of the fixed gates. Other ideas involve a

combination of engineering and socio-economic reorganisation, notably the proposal to move the passenger port to outside the Lido inlet, which would allow greater flexibility in re-dimensioning the inlets.

ISN'T THERE A BETTER WAY OF SPENDING ALL THIS MONEY?

Some opponents argue that the money spent on the barriers would be better spent on the other diffuse and restoration measures described in the last chapter, and that these 'alternative' measures will be enough to protect the city. However, careful analyses and computer simulations show that the diffuse measures would not have a significant effect on high waters, nor would their combined effect reduce levels by more than a few centimetres. They certainly could not provide the protection needed to cope with a storm surge like that of 1966.

While not all scientists are confident that the radical design chosen for the barriers is the right one, most are convinced that, if flood protection is the immediate priority, then flood barriers must be built now. For them, the barrier is the only way of protecting the city and lagoon from future floods. Alongside this, the lagoon will continue to become a bay despite the barriers – so other works are needed to combat the erosion and associated ecological problems.

left A prototype single gate was tested near the Lido inlet – many people feared that these red towers were a permanent feature, and there were huge protests when the prototype was first seen in Venice

A summary of interventions

The scale of the projects underway to solve the 'Venice problem' is staggering. Their development needs careful coordination, as each will affect the other

HIGH waters and tides, erosion and pollution, physical and socio-economic deterioration — the many risks and problems that face the Venice lagoon are intricately bound together. While each requires its own tailored solution, the problem can only be solved if it is tackled as a whole.

Scientists and engineers are dealing with many layers of uncertainty about the semi-natural system and the measures proposed to modify it. There is much that is not known, much to study, and each element of the chosen approach must remain flexible, ready to respond and adapt to new knowledge. Meanwhile, the mobile barrier and other interventions are proceeding, from feasibility to planning and final execution, always under constant review by all parties.

The success of the interventions will be measured not just in technical terms, but also in terms of the wishes of many others. These include people who live and work in the city, those who depend on the lagoon's industry and port, those who visit the city and those who treasure the lagoon's ecosystems. All have a stake in Venice's future.

Few projects, least of all those that involve complex environmental systems, have the benefit of perfect certainty and complete knowledge, but decisions still need to be made and acted on. This is especially so for large-scale projects. But such projects do provide an opportunity to learn: from actual impacts as well as during construction, at different phases before the project is completed. Protecting the city from flooding is urgent. The barrier is not the final solution, but it buys the city time.

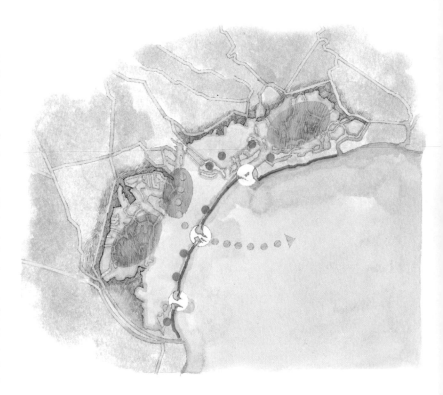

 Exceptional high tides (over 110cm)
Mobile flood barriers at lagoon inlets

Deteriorating jetties
Reconstruction of jetties

 Medium to high tides (to 110cm)
Local interventions to protect urban areas and embankments

Physical deterioration of lagoon
Restoration, dredging and rebuilding features

Access closed to tidal flow
Re-opening of fish farms and other isolated areas

 Coastal erosion
Reinforcement, restoration of walls and beaches

 Environmental damage and pollution
Restoration and prevention

 Oil tanker traffic pollution risks
New terminal proposed for offshore

"If I had to choose between the Port or Venice, I would choose Venice"

Claudio Boniccioli, former President of the Venice Port Authority

CHAPTER 5 SUMMARY

- ■ To protect against extreme flooding, construction has begun for the **MOSE** scheme of mobile barriers and breakwaters at the lagoon inlets
- ■ Further measures are needed to reverse the degradation of the lagoon system, to complement and mitigate the negative impacts of the barrier system
- ■ Venice can benefit from experience gained in reducing flood risk in other places, notably flexibility during implementation, institutional rigour and unexpected benefits/adaptations of the system
- ■ This is a new phase for science in Venice, where some old questions are sharpened by implementation of the barriers and many new questions arise
- ■ The long-term options for Venice will need to be considered in the context of uncertainty over future global environmental change

Policy-makers aspire to the adoption of innovative, sophisticated engineering solutions alongside the preservation of traditional lagoon activities

Futures

*"A future that we would like
to think has no end"*

Bruno Dolcetta, President, Insula SpA

TO most people, a world without Venice is unthinkable, but as the previous chapters have shown, saving the city is a daunting task. It needs the concerted efforts of everyone – politicians, scientists and engineers, every citizen and the international community – to succeed. There are many possible futures for Venice, serving different interests and rival demands, which must be made explicit, prioritised and reconciled. Good science is the bedrock for any lasting solution, but in the end it is up to the people who live in Venice – and also all who love it – to answer the question, 'what sort of Venice do we want?'

75

THIS FINAL CHAPTER LOOKS AT
- **THE UNCERTAINTIES THAT FACE THE DECISION-MAKERS**
- **THE ROLE OF SCIENCE IN MOVING FORWARDS**
- **WHAT THE WORLD CAN LEARN FROM VENICE**
- **WHAT CHOICES ARE AVAILABLE**

left St Mark's Piazzetta, with the landmark statues of St Mark and St Theodore, flanks the open waters

Venice and global warming

Climate change may pose the most serious threat to the long-term survival of Venice as sea level continues to rise

THERE is substantial agreement among international scientists that human-induced climate change is a reality, and that it will lead to increases in average sea level and changing weather patterns. Currently available estimates of future changes are mainly global averages and, to date, there is little agreement on the rate and scale of sea-level rise on a regional or local level. In addition to sea level change, regional weather patterns are being affected and in Venice this could mean changes to the intensity and frequency of winds, surges, rainfall and indeed extreme events.

These uncertainties are bound to continue but may be reduced as we develop a better understanding of how the world's climate systems work, based on detailed studies of present climate, analyses of past climate conditions and improved simulation of future scenarios. The association between observed, measured changes and human-induced factors is still tenuous – because climate and sea level are both inherently variable and because there is a time lag between cause and effect of climate processes. We have less than a hundred years of measurements since industrial activities and pollution levels reached a globally significant level.

Climate scientists liaising through the United Nations' International Panel on Climate Change (IPCC) show that global sea level rise could be anywhere between eight and 88cm by 2100. The IPCC is now making a special effort to establish the likelihood of the various levels within that wide range of possibilities. Their predictions are based on data up to 1998, but more recent measurements of glacial melting already question some of the scientific assumptions used in the general models. Others challenge the economic assumptions about industrial development and the associated amount of atmospheric emissions. So the range of possibilities to plan for is very wide.

Sea level trends are particularly hard to compare due to lack of uniformity in tide gauge systems around the world and the fact that most gauges are in coastal regions (and therefore affected by local variables). Satellites have recently become available for measuring sea level, especially in the open sea, which is extremely useful. Readings for the Mediterranean are available for the past eight years and show a rise of 2.2mm/year overall – in the eastern Mediterranean, however, the rise is as much as 9.3mm/year while in the Ionian it is *falling* by 11.9mm. It has remained more or less level in the western Mediterranean and in the Adriatic. But even if the trends are unclear and the signals they give are conflicting, the growing understanding of how the Earth's climate system works and interacts with the oceans point in the direction of unprecedented rates of change and a greater chance of unpredictable catastrophic weather-related events. Oceans and ice sheets are so vast, they are still reacting to climate changes that occurred in a more distant past; researchers have also shown how rain, wind and atmospheric pressure can influence relative sea level by as much as +/- 1m by their natural variations.

LOCAL UNCERTAINTIES

At a local level, uncertainty is even greater, although the positive link between average sea level and flooding frequency has already been made (see Chapter 3). The possible sea level rise

"Sea level rise is like a rope that is tightening around the neck of Venice"

Alberto Tomasin,
Dept of Applied
Mathematics,
Università
Ca' Foscari, Venice

Storm waves overcome the sea walls during the flood of 1966. Will Venice suffer more, or fewer, storms in the coming years?

left St Mark's Piazzetta became part of the open lagoon during the 1966 flood

for Venice is in the region of 12.6–70cm by 2100, according to a broad literature review. Besides the general sea level rise and the continuing role of subsidence, it is imperative to better understand the local conditions that affect not only sea level rise but weather patterns. Changes in the frequency and magnitude of storm surges that cause flooding in Venice is an expected consequence of climate change – but in Venice no trend has emerged yet. Storms could become more frequent and/or stronger, or fewer but more intense.

Associated with this, Venice is constantly improving its ability to predict high waters with greater accuracy on timescales of hours and days. Two models are used – one based on a *statistical* approach, which means that the model tells its operator the likely water level in response to data from weather systems moving in the direction of

the city, based on the historical correlation of certain winds and observed water levels in Venice. The other approach is *deterministic* – by setting up a model of what is known about the many parameters affecting tide levels, data for each of these is fed into the model, which then calculates the expected water level on the basis of known interrelationships between the parameters. These systems are crucial to the success of the barrier management scheme.

WIDER CONCERNS

Mobile barriers may be able to protect Venice from extreme floods, but much of the northern Adriatic coastline is vulnerable to permanent flooding. Venice is enormously privileged – less famous places, lacking its high international profile, will probably not be helped.

Science and uncertainty

In predicting and adapting to the future of the lagoon we have to accept uncertainty as a reality that must be managed

IT is unrealistic to expect science to provide simple, clear answers about such a complex, dynamic system as the Venice lagoon – partly because scientists do not know enough about how the lagoon system works. There is a huge amount of work to do to understand key physical dynamics, how various physical and biological components interact, let alone to predict how their relationships will change once the barrier and other restoration and defence elements are in place. Each change will ripple through the system, affecting all the other factors. Adaptation takes time.

The challenge is an exciting one, with implications well beyond Venice, and well beyond science. The Venice lagoon is a natural laboratory for studying the impacts of global, as well as local, changes on a coastal ecosystem and the lives of its people. Even after more than 35 years of intensive field research, theoretical advances and implementation of extensive safeguarding measures, the open questions regarding the Venice lagoon system have changed little – as our knowledge deepens, so does the detail required.

THINKING BEYOND THE BARRIER
The barriers are designed to protect Venice from flood surges of more than 110cm above the reference water level. The calculated closure scenarios incorporate a sea level rise of 16–25cm, which would bring the expected number of annual closures from seven to 39 times. The barrier scheme's Environmental Impact Study stated that a 50cm increase in sea level was not considered, as it would be catastrophic not just for Venice but for the entire world. The barrier designer's cost-benefit analysis gives the project an expected lifetime of 100 years or less. Thus the mobile barriers are not a complete or final solution, but they could give Venice some more time. If all goes well, and the fears for the lagoon's ecology are not realised, the barrier will not significantly alter the basic character of the lagoon during its operating years. But the city will continue to sink and water levels will continue to rise.

RADICAL IDEAS?
In the 1970s a project investigated the viability of raising Venice, building by building, out of harm's way – by injecting materials underground to push it upwards. A trial on one of the small islands in the lagoon, Poveglia, revealed the heterogeneously distributed and highly variable ground layers beneath Venice and the instabilities that could be caused by differing responses to the injections within a single building or even adjacent ones. Recently the concept has re-emerged in a far more sophisticated and technologically advanced form.

The injection of sea water is being proposed at a depth of 600–800m via a series of wells placed in a 5-km radius around Venice. The aim, this time, is to uniformly raise a large area of the lagoon and the potential is a gain in height of 30cm relative to mean sea level. It will take a few more years to

"If there is one thing that researchers do agree on, it is that the gates are not the sole solution."

The Economist,
September 2003

far left Much of the Po Plain and other areas on the mainland in the Venice region are nearly at sea level or below it

investigate the feasibility in greater detail – particularly as the geological composition below Venice and its lagoon and distribution of different types of soil down to nearly 1km from the surface has presently only been characterised in a couple of points. Even in the event of promising indications, the technique must first be tested in a pilot well. If it works, Venice would have several more decades in its fight against sea level rise and increasingly frequent 'chronic' flooding.

Alternatively, Venice could look to the Dutch example in the Eastern Scheldt, where an area was permanently closed off from the sea and transformed from a saline system to freshwater. Aside from the biodiversity implications of losing many plant, invertebrate, fish and bird species, pollution control would be a major issue – including a reduction in contaminating inputs from the watershed. Equally essential would be the implementation of a complete sewage collection and treatment system for the historical centre and other inhabited areas of the lagoon.

At the opposite extreme is the option to accept the forces of nature and let the lagoon develop naturally into a marine bay, creating other, different habitats, and essentially relinquishing the city to the waves.

"The lagoon system and its dynamics represent the perfect laboratory to develop an approach in which scientific, technological and economic factors interact and influence each other. If we can find a solution to our problems in Venice then it might be possible to apply that solution elsewhere in the world"

Pierpaolo Campostrini, Director, CORILA

What future does Venice want?

Tackling complex environmental problems is not only a matter of finding scientific and technical solutions. It is also important that people have the political will to solve them, and the financial means to do so

IN nature, lagoons are ephemeral, but in the history of Venice, man has actively maintained the stability of the system and its 'natural' dynamics. The environment and culture of Venice and its lagoon have co-evolved and maintained each other for more than ten centuries, and they will always have to be managed in some way. The question is, how is it best managed, and for whom?

Venice excites huge emotions. Each interest group sees the city and lagoon in different ways, valuing some attributes above others. It is perfectly possible to make deliberate choices about the direction to go in – whether, at its most radical, to return the lagoon to the sea (or to land), letting nature finally take its course, or to take the necessary measures to maintain the precarious balance that exists today.

The various institutional roles are outlined in the Appendix – the Venetian district is characterised by a large number of authorities with overlapping competences (many of which possess the power of veto) resulting in a substantially weakened planning system with divided authority and restricted impact. Scientists therefore have an especially important role. By sharing data and working across disciplines, they can be called upon to clarify the present situation and future scenarios so that it is possible to make deliberate choices in the face of competing demands on the delicate city and fragile lagoon.

WHAT IS VENICE BEING SAVED FOR?

FOR NATURE? Any consideration of the future of Venice must guarantee the conservation of this unique system. Revitalising the lagoon's physical structure and associated biological communities is essential. Further work is necessary to properly appreciate the potential of lagoon-wide measures to mitigate water levels and the impact of these (both positive and negative) on other aspects of the lagoon's ecological balance. On the global stage, too, the preservation of endangered wetlands is an

international priority of increasing urgency.

But what sort of restoration is acceptable, and to whom? The new projects are not returning the system to its prior state, but creating a different balance of nature (including the increase of some aggressive grasses and bird species that are more characteristic of the mainland). It must be recognised that there is no single reference condition to work towards in rehabilitating the natural environment and dynamics of the lagoon. Interaction over centuries of human intervention, along with the relative influences of rivers and the sea, have produced a mosaic of differing geographical and functional circumstances, each requiring its own restoration criteria.

FOR VENETIANS? With worsening flooding, structural decay and pollution, matched by increasing costs of living and buildings maintenance – problems exacerbated by market distortions caused by mass tourism – Venetians are deserting their city. How will they be persuaded to return? Repairing the city fabric, making it fit for the 21st century – the essential infrastructural works, restoring abandoned buildings – could all make it worthwhile. Policy measures are needed to stimulate traditional as well as post-industrial productive activities and reduce the concentration on tourism monoculture. Aside from the obvious socio-economic effects of the 'international mass consumption' of Venice, tourism damages the already precarious urban and natural environment. The effects (intense boat traffic, litter) are evident along the main thoroughfares of the city and are beginning to spill over into the lesser-known more fragile areas, such as the northern lagoon. The lagoon, too, is a place of work for Venetians – as the waters return to health, so can fishing be competitively maintained, balancing economic demands with the needs of the ecosystem for long-term sustainability. But tourism, along with the port, is Venice's principal earner so any adjustment needs to made

"Venice is one of the most consistent and continuous examples of permanent urban planning affecting both settlement and environment known to humanity"

Territorial Plan for Conservation and Development of the Venetian Environment and Settlement System (1995)

above, left to right
San Giobbe housing project. A Venice in Peril Fund initiative to demonstrate the feasibility of traditional building techniques to restore popular housing in Venice.
Marghera industrial complex.
Venice full of tourists
right Sea walls at the Lido

carefully. It is also important for Venice to settle its role within the context of the 'metropolitan city', including Mestre and other associated mainland settlements.

FOR TRADE, INDUSTRY AND AGRICULTURE?

Each has been targeted as villains in Venice's story, and there are increasingly tight restrictions on their waste, run-off and emissions. Time bombs from the past still rest in the sediments of the lagoon as well as the many hectares of abandoned industrial areas earmarked for remediation measures in Marghera. But the industries that cluster round the port are vital parts of the economy, bringing wealth to the region and employing thousands of people. To what extent must Venice learn to live alongside these economically indispensable sectors? How much can the lagoon absorb without further damage? Would an offshore oil terminal reduce the risk of tanker pollution? Can alterations be made to the burgeoning growth in passenger traffic aboard gigantic cruise liners that pass right through the heart of the city and deliver yet more 'bite and run' tourists, as they are now called?

FOR THE WORLD? Venice is a UNESCO World Heritage Site: people all over the world feel they have a stake in its future, too. But does that mean Venice is obliged to allow itself to be consumed by tourism, and are tourists merely supporting the preservation of a fossil? A review of the criteria that put Venice on the list of World Heritage Sites carries some important reminders – in addition to Venice's unmatched artistic, architectural and cultural development and influence around the world, it is an outstanding example of wetland biodiversity and "symbolises the victorious struggle of mankind against the elements, and the mastery men and women have imposed upon hostile nature".

Venice has the potential to be a powerful case study in sustainable social and economic development, leading the world in learning to live with natural systems in more harmonious ways, rather than trying to dominate them. The scientific community recognises that the answers they can provide are only part of the final equation. But theirs remains a vital voice.

"It is a delicate task to get people to work together at the bottom of the pyramid, where the real work gets done!"

Dr Roberto Frassetto, founder CNR Institute for the Study of Large Masses

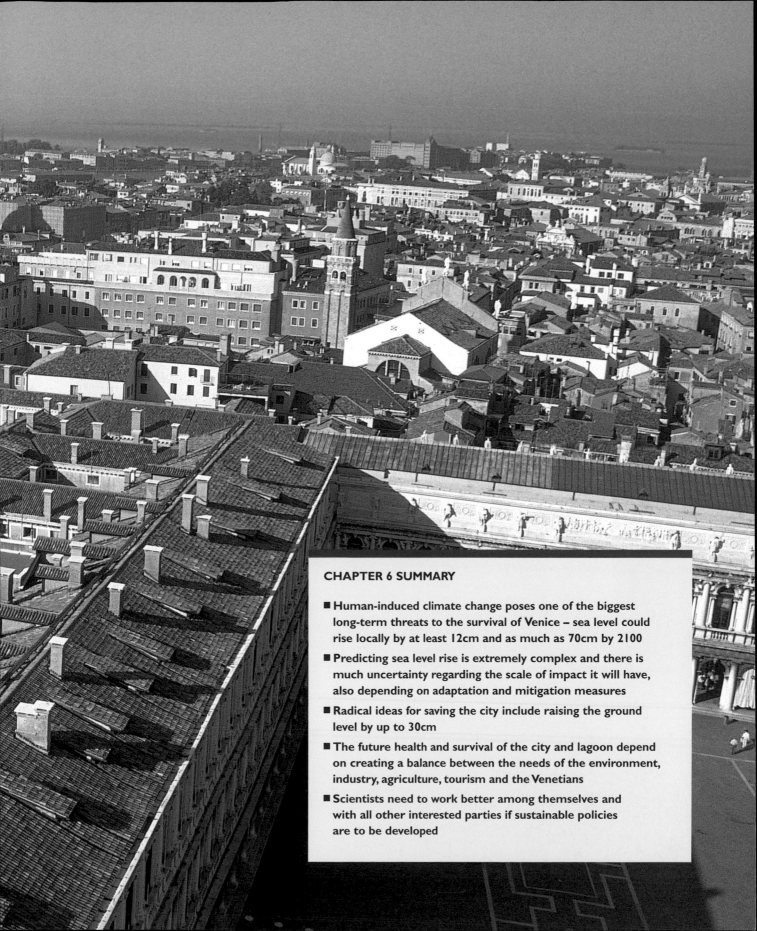

CHAPTER 6 SUMMARY

- Human-induced climate change poses one of the biggest long-term threats to the survival of Venice – sea level could rise locally by at least 12cm and as much as 70cm by 2100

- Predicting sea level rise is extremely complex and there is much uncertainty regarding the scale of impact it will have, also depending on adaptation and mitigation measures

- Radical ideas for saving the city include raising the ground level by up to 30cm

- The future health and survival of the city and lagoon depend on creating a balance between the needs of the environment, industry, agriculture, tourism and the Venetians

- Scientists need to work better among themselves and with all other interested parties if sustainable policies are to be developed

Appendix Authorities, organisations and legal background

SPECIAL LAWS FOR VENICE

During Venice's time as a Republic (607–1797), there were carefully specified and vigilantly applied laws for the lagoon and mainland domains. This attention dwindled from the 19th century onwards and consequently the lagoon's defining characteristics and underlying dynamics altered significantly. Only following the 1966 disaster was care taken again for the special characteristics of the lagoon and its inseparability from safeguarding measures for the unique city.

While ordinary laws set the duties of the State and local authorities as regards territorial management, the Special Law of 1973 established that Venice is of pre-eminent national interest and gave the State authority to determine specific objectives and associated financing for measures. The main objectives of the first Special Law were:

■ Safeguarding of the environment (landscape, historical, archaeological, artistic features);
■ Protection of the hydraulic and hydrogeological equilibrium;
■ Regulation of watercourses (natural and artificial) feeding into the lagoon;
■ Reduction, followed by regulation, of tide levels;
■ Coastal defence works;
■ Pollution protection works.

Another iteration of the Special Law in 1984 had a slightly different orientation. It committed the State to the design, experimentation and execution of works to:

■ Re-establish the hydrogeological equilibrium of the lagoon;
■ Arrest and invert the process of degradation;
■ Protect the insulae;
■ Protect the urban settlements from the exceptionally high tides, "also via measures at the inlets (...), with the characteristics of experimentability, reversibility and gradualness".

The law also allowed the Public Works Ministry to identify a single operating agency to guarantee the "unitary nature" of the interventions for safeguarding the lagoon. Consorzio Venezia Nuova, made up of a group of Italian construction and engineering firms, was nominated as the executive body – essentially a private company with public objectives. The 1984 Law also established the Comitatone (see below) to oversee the implementation of safeguarding objectives and budgetary allocations

among the various institutional bodies. Then in 1992 the Third Special Law made it a requirement to obtain the opinions of the Veneto Regional administration and those of the Venice and Chioggia municipalities for safeguarding strategies and measures. Its main other points are:

■ Adjustment and reinforcement of the long breakwaters at the three lagoon inlets;
■ Local defences from high waters for built areas;
■ Restoration of lagoon morphology;
■ Halting the process of the lagoon's deterioration;
■ Coastal defences;
■ Substitution of petrol tanker traffic in the lagoon;
■ Opening up of the closed acquaculture areas to tidal expansion.

As well as setting out the safeguarding objectives, each Special Law also set aside State funds for fulfilling them. Since there has not been a Special Law since 1992, Parliament has meanwhile set aside funds for safeguarding Venice in its annual budget. But in 2003, for the first time, no provisions were made for Venice in the budget, nor again in 2004. Funding for the mobile barriers, however, has continued via the Strategic Objectives, a measure introduced to fund major infrastructural projects such as this and a bridge from the Italian mainland to Messina, in Sicily.

INSTITUTIONAL FRAMEWORK

Management of Venice and its lagoon is notoriously complex. Indeed, the lack of progress on implementation of a robust plan for safeguarding measures has been attributed to overlapping and fragmented institutional responsibilities among various administrations and even within departments of a single entity. Venice is the capital of the Veneto Region in northern Italy and of Venezia Province, one of the region's seven provinces. National ministries and regional, provincial, and local authorities, as well as the Venice City Council, are all responsible for various aspects of the administration of the city of Venice, its lagoon, and the surrounding areas. For example, the land that drains into the Venice Lagoon, an area of almost 1,900 km^2 is administered by 100 local authorities.

The 1973 Special Laws assigned specific responsibilities to various authorities concerned with safeguarding Venice and the lagoon. The Italian State has responsibility for the physical

safeguarding and restoration of the hydrogeological balance in the lagoon. The Veneto Region is responsible for the abatement of pollution, especially from the drainage basin. And the city councils of Venice and the town of Chioggia at the southern end of the lagoon are responsible for urban conservation and maintenance, as well as activities aimed at promoting socio-economic development.

The Venice Water Authority (Magistrato alle Acque), which was founded in 1501, has responsibility under the Italian Ministry of Public Works for ensuring the survival of Venice, its lagoon, and the living species that inhabit it, and for protecting the lagoon from both natural events and human impact.

In 1992, a general plan backed by law to implement measures designed to safeguard Venice was launched by the Venice Water Authority and the New Venice Consortium. The plan is two-pronged. The first prong relates to the protection of the city from high water levels and flood damage by improving its physical defences by, for example, raising embankments, and also by reinforcing the coastline to protect it from storm damage.

The second prong aims at improving the environmental quality of the lagoon through a series of interventions that tackle the factors causing the deterioration of the lagoon's ecosystem, and that also reconstruct and maintain the natural environment of the lagoon.

■ Commissione di Salvaguardia – Safeguarding Commission for Venice

Instituted by the First Special Law, this committee was granted authority to express its discretionary opinion on all building works as well as territorial transformations and modifications planned by private and public bodies everywhere within the Venice lagoon boundary. There are about 20 members, including UNESCO and the National Research Council (CNR), presided over by the President of the Region. To some extent it overlaps with the Town Council's land use planning remit and it has been accused of causing bureaucratic delays. Decision-making parameters are dominated by aesthetic rather than technical considerations.

■ Comitatone

The second Special Law (1984) instituted a mixed committee of government ministers and local authorities, known as the Large Committee or *Comitatone*. It decides strategy, coordination and control of the implementation of all measures to safeguard Venice and the lagoon,

The Science of Saving Venice

especially how to divide the budget. The committee is chaired by the President of the Council of Ministers (the Italian Prime Minister) and consists of the heads of five ministries, their executive branches and the various local administrations, including the President of the Venice Water Authority (Secretary), the Minister for Infrastructure and Transport, the Minister of the Environment, the Minister of Cultural Heritage, the Minister of Transport and Navigation, the Minister of Universities and Scientific and Technological Research, the President of the Veneto Region, the Mayor of Venice, the Mayor of Chioggia and two representatives of the many other local authorities bordering the lagoon.

■ Magistrato alle Acque (MAV) – Venice Water Authority
www.magisacque.it

The Venice Water Authority is a technical agency of the Ministry for Infrastructure and Transport with direct and primary responsibility for the safeguarding, security and hydraulic protection of a large area spread across a number of regions (Veneto, Friuli and Lombardy). It was established in 1907 but the name dates back to an organ of the Venetian Republic of the same name, founded in 1501.

MAV duties cover five national river catchments (Adige, Brenta-Bacchiglione, Piave, Livenza, Tagliamento), one international catchment area (Isonzo), and the lagoons of Venice, Marano and Grado. With regard to the Venice lagoon, MAV carries out systematic monitoring activites and oversees the planning and execution of safeguarding measures via Consorzio Venezia Nuova, its executive agency. A few technical offices have remained within MAV, notably the Anti-pollution Service, which publishes a monthly digest of lagoon water quality according to a series of physical-chemical characteristics at five monitoring stations.

■ Consorzio Venezia Nuova (CVN)
www.salve.it

The Consorzio Venezia Nuova is the executive agency of the Ministry for Infrastructure and Transport – Venice Water Authority (as established by the second Special Law) responsible for planning and implementing the measures to safeguard Venice and its lagoon, delegated by the law to the State.

The second Special Law established that a single body could take on responsibility for the interventions as a whole, on behalf of and controlled by the Venice Water Authority, and on the basis of a general plan of interventions defined and approved by the *Comitatone* (and therefore by the institutions represented on the committee) and by Parliament. Hence CVN is required to prepare and implement an 'integrated plan' to tackle the various aspects of the physical and environmental safeguarding of the lagoon ecosystem in a unitary and organic fashion and with a systematic approach.

Measures to monitor and improve the quality of water and sediment in the lagoon – most recently known as the MELa Programmes 1 and 2 – are also the responsibility of the State and carried out by CVN, in collaboration with the Regione, ARPAV and university research departments.

■ Regione Veneto – Regional administration
www.regione.veneto.it

Pollution abatement within the lagoon watershed area is the responsibility of the Veneto Region, which has established a framework programme of measures to monitor and reduce pollution in the drainage basin, known as the Master Plan 2000. Pollution abatement and prevention measures programmed by the Veneto Region for the drainage basin to reduce the loads arriving in the lagoon are now closely coordinated with lagoon water quality control measures (the remit of MAV/CVN).

The Regione is also responsible for drawing up the territorial plan for the conservation and development of the Venice environmental and settlement system (PALAV). It covers 16 municipalities from three provinces (Padua, Treviso and Venice). It was requested under the 1973 Special Law – it took over a decade to formulate, underwent several iterations and was finally approved in 1995.

■ Provincia di Venezia – Provincial administration
www.provincia.venezia.it

The Provincial Administration does not participate in the *Comitatone*, except on a consultative basis, although it oversees certain key elements of Venice's natural resources, along with about 40 other municipalities within its domain. It has some land use management and environmental protection responsibilities, notably regulating fishing activities in the lagoon as well as hunting licences and it runs the environmental police department. It oversees the protection of flora and fauna, natural parks, organises waste disposal at the provincial level and controls effluents and emissions and noise pollution.

It is responsible for certain restructurings and restorations financed by the Special Law (notably schools and the island of San Servolo); territorial planning including landscape protection and environmental resources; planning and environmental education for pollution prevention and control. A small area of the Venice lagoon falls within the domain of the Province of Padua.

■ Comune di Venezia – Venice Municipality
www.comune.venezia.it

The Venice Municipality governs the main urban areas of the lagoon – the historic centre of Venice, Lido, the islands of Murano, Burano, Mazzorbo and Torcello, and the coastal strip of Pellestrina and San Pietro in Volta in the southern lagoon, as well as Mestre and Marghera. (About nine other municipalities have territories within the lagoon and are duly subjected to the special legislation for Venice.)

In the context of the Special Law, the *Comune* must look after Venice's historical, cultural, architectural and environmental heritage while also addressing the socio-economic factors that determine the city's identity and wellbeing – everything from stimulating productive activities and controlling tourism to assisting young couples with obtaining affordable housing and providing crèches for babies. Specifically, Special Law funds are used by the Town Council for:

● Acquisition, restoration and refurbishment of buildings for residential, social and cultural, commercial and artisan uses – considered essential to maintaining the socio-economic identity of lagoon settlements;
● Basic infrastructure (street lights, utilities etc,) as well as bridges, embankments and canals within the municipal domain, i.e. the 45km inner canal network (whereas the larger canals are within the MAV remit along with the rest of the lagoon open waters);
● Subsidies for restoration and maintenance works for private buildings, including for example the installation of lifts in tall palaces (pending approval of the Safeguarding Commission and the local branch of the cultural heritage ministry);
● Acquisition of areas to be converted to productive activities and the associated infrastructural requirements.

The *Comune* set up a mixed company – Insula spA – for the integrated management of canal dredging, raising street levels, revision of utility pipelines, underground maintenance works, etc. (see below). The tide forecasting office (CSPM – see below) is also part of the *Comune*. Executives from the *Comune* set up a working group in 1999 to review the Environmental Impact Study of the mobile barriers.

Total State funds allocated via the Special Law (1984–2000): nearly €6 billion

Public Works Ministry-CVN €2.2bn
Others* €0.5bn
Comune di Chioggia €0.2bn
Comune di Venezia €1.4bn
Regione €1.4bn

* of which €17m has been used for research via the Ministry for Education, Universities and Research.

OTHER ORGANISATIONS

■ Agenzia Regionale per la Prevenzione e Protezione Ambientale del Veneto (ARPAV) – Veneto Regional Agency for Prevention and Environmental Protection
www.arpa.veneto.it
The Veneto Regional Agency for Environmental Protection (ARPAV) is essentially the technical branch of the regional administration. Its tasks can be summarised as:
* Environmental protection and monitoring;
* Weather forecasting, monitoring and statistical elaborations;
* Organisation and management of the regional information system for environmental monitoring and environment related epidemiology;
* Environmental education and information services;
* Technical and scientific services for environmental impact assessments and evaluation of environmental damage.

■ Agenzia per la protezione dell'ambiente e per i servizi tecnici (APAT) – National agency for environmental protection and technical services
www.apatvenezia.it
The Hydrographic Office has been operating in the Lagoon since 1907 within Magistrato alle Acque but recently changed to be run from central government. It manages a network of 52 tide gauge stations for the systematic measurement of tide level and related parameters, such as wind direction and wind speed, atmospheric pressure, precipitation, and wave-height.

■ Consorzio per la gestione del Centro di Coordinamento delle Attività di Ricerca inerenti il Sistema Lagunare di Venezia (CORILA) – Consortium for Coordination of Research Activities Concerning the Venice Lagoon System
www.corila.it
CORILA is an association of Ca'Foscari University and the University Institute of Architecture of Venice, the University of Padua and Italy's National Research Council. A non-profit organisation, it is overseen by the Ministry of Education, Universities and Research (MIUR). It was founded in 1999 to coordinate and manage research on the Venice Lagoon and thereby provide decision support information to policy makers and public administrations dealing with Venice. Activities are organised within the framework of three-year research programmes, which in turn are divided into thematic areas and research lines. The principal thematic areas are: economics, architecture and cultural

heritage; environmental processes; data management and dissemination. The first research programme (2000–2003) costed approximately €10,8 million of which nearly 60 per cent was funded by the Special Law via the Ministry for Research. Co-financing was provided by other administrations as well as the research departments and other partners themselves.

The second research programme (2004–2007) has just under €6m from the Special Law plus co-financing. Accordingly, CORILA facilitates the acquisition of knowledge and information information on the physical system, territorial, environmental, economic and social aspects of the lagoon and lagoon settlements; processes and manages this information in an integrated framework; carries out interdisciplinary scientific research projects pertinent to the problems of the Venice lagoon; facilitates interaction with the international scientific community. The four partners of CORILA have naturally been active participants in research on the Venice lagoon system and the city itself since the earliest times.

■ Centro Previsioni e Segnalazioni Maree – Tidal Forecasting and Early Warning Centre
www.comune.venezia.it/maree/
Founded in 1980, this organ within the Venice Municipality is responsible for the study and forecasting of storm surge events and alerting the city in case of flood events. Observation of sea level and meteorological parameters is carried out through a monitoring network (11 stations), that gives a real-time view of marine and weather conditions in the Venice lagoon and along the Adriatic coast. All stations measure sea level and some also collect meteorological parameters: air pressure, humidity, wind velocity and direction, waves and air temperature.

■ Consiglio Nazionale delle Ricerche (CNR) – National Research Council
www.cnr.it
The Italian National Research Council is a high profile public organisation in the field of scientific and technological research. Founded in 1923, it is composed of dozens of separate institutes, each with a particular specialisation. In response to the increasing worldwide concern for the survival of Venice, CNR established the Institute for the Study of the Dynamics of Large Masses in 1969, now incorporated within the Marine Sciences Institute (ISMAR). Created first as a laboratory, it has spread from basic research in oceanography and geology to applied research. The other Venice-based branch of CNR-ISMAR is the Institute for Marine Biology, established in 1946 as the National Centre of Talassographic Studies, which focuses on pure and applied biological oceanography, marine and lagoon biology.

■ Consorzio per la Ricerca e la Formazione (COSES) – Consortium for Research and Training
www.provincia.venezia.it/coses/
COSES was established in 1967 by the municipal and provincial administrations to carry out analyses, studies and projects to support public sector activities – essentially via market research, data collection and statistical analyses concerning the urban and regional economy, building sector and housing, commercial distribution, tourism, culture, teaching and education, immigration, demographics, transport (especially water) and urban planning.

■ Insula SpA
www.insula.it
Founded in 1997 by Venice Town Council (52 percent share) together with Vesta (waste management), Enel.Hydro (electricity), Italgas (gas) and Telecom Italia (telephones), Insula's mission concerns urban maintenance and, more precisely, measures such as clearing canals of accumulated silt, restoration of canal walls, foundations and façades of buildings lining canals, restoring bridges, rationalisation of urban subsoil (utility lines and sewer system), maintenance and renovation of paving, raising of footpaths above the level of medium-high tides (local protection). Project integration and works coordination (all parties involved in the operations work side by side, not least the public utilities, who are Insula's partners) is therefore essential to this complicated process to minimise inconvenience, while also boosting the efficiency, in terms of economies of scale in such a delicate urban environment.

■ Istituto Veneto di Scienze Lettere ed Arti
www.istitutoveneto.it
The Istituto Veneto di Scienze, Lettere ed Arti was founded by Napoleon Bonaparte "to collect discoveries, and to perfect the arts and sciences". Its current mission is to increase, promulgate and safeguard the sciences, literature and arts, bringing together outstanding figures from the world of scholarship. The Institute also supports special research projects that concern Venice and the Veneto, and which are addressed at the international community. Together with various universities and the National Research Council, it has also set up specialised centres for research into environmental questions, into philological and literary aspects of the language of the Veneto, and into Hydrology, Meteorology and Climatology. It runs a programme to integrate and share environmental data among all the major institutions and research bodies.

■ Università Ca' Foscari di Venezia
www.unive.it
With four faculties and 19 departments, the University dates back to the late 19th century

and its students account for a significant proportion of the local population. It covers many areas of chemistry and environmental sciences; in the area of economics it carries out specific environmental economics studies; in the area of mathematics and IT it has developed and has access to models and data management instruments; significant contributions are also made in the field of law.

■ Università IUAV di Venezia – IUAV Venice University

www.iuav.it

Founded in 1926 it is an international reference point for architecture, history, design, and restoration as well as for town and land use planning. It also has laboratories for construction science and analysis of ancient materials. There are about 8,000 students enrolled, 209 tenured professors and 240 contract professors plus support and technical personnel.

■ Università degli Studi di Padova – Padua University

www.unipd.it

Founded in 1222, it was the first university in the world to award a degree to a woman, Elena Lucrezia Cornaro Piscopia, in 1678 (in philosophy). It has 13 faculties and 62 departments and is a world leader in hydraulic engineering, biology, agricultural sciences, chemistry, mathematics and many other branches of science – not to mention centuries of tradition in law and medicine – matched by first-class facilities and instrumentation.

■ UNESCO Office Venice – Regional Bureau for Science in Europe

www.unesco.org

Following the disastrous floods of 1966 in Venice and Florence and the Italian Government's invitation for UNESCO to contribute, the Liaison Office for the Safeguarding of Venice was established in 1973 on the occasion of the UNESCO International Campaign for the Safeguarding of Venice. In 1988 the UNESCO Scientific Co-operation Bureau for Europe (SC/BSE) was relocated from Paris to Venice and renamed as Regional Office for Science & Technology for Europe (ROSTE). In 2002, UNESCO established a single office in Venice with the mandate to achieve UNESCO's and Member States' goals in the fields of science and culture. The UNESCO Office Venice actively promotes, sponsors and convenes international scientific and cultural events in Europe and in the Mediterranean region. A unifying theme for UNESCO is its contributing to peace and human development in an era of globalisation, through education, the sciences, culture & communication.

● International Private Committees for the Safeguarding of Venice

Following the appeal launched by the Director General of UNESCO in 1966, over 50 private organisations were established in a number of countries to collect and channel contributions to restore and preserve Venice. Over the years, the International Private Committees have worked closely with the Superintendencies of Monuments and Galleries of Venice, through UNESCO, to identify and address priority needs. Since 1969, they have funded the restoration of more than 100 monuments and 1,000 works of art, provided laboratory equipment and scientific expertise, sponsored research and publications and awarded innumerable grants for craftsmen, restorers and conservators to attend specialist courses in Venice. Expenditure by the Private Committees for the five-year period 1996–2000 was well in excess of 5 million (about £3 million).

■ Venice in Peril

www.veniceinperil.org

The instigators and funders of the Cambridge Research Project. Venice in Peril is a registered British charity that works for the safeguarding of Venice, by restoration and by research into its ecological problems. It was founded after the great flood of 1966 and since then has restored over 25 monuments and works of art in the city. It works as part of the international association of private committees for Venice, which has the status of a Non-Governmental Organisation in operational relations with UNESCO.

THE CAMBRIDGE PROJECT

"Flooding and Environmental Challenges for Venice and its Lagoon – State of Knowledge" (2001–2004)

The mission is to promote the objective study and review of information concerning key aspects of the flooding and environmental issues relevant to Venice, in an international dimension. This project was set up at the instigation of and with financing from the Venice in Peril Fund and run by Cambridge University (Churchill College and Cambridge Coastal Research Unit) with local support from CORILA. A series of preparatory workshops were held in September 2002 and a four-day International Discussion Meeting was held at Churchill College, Cambridge, in September 2003. The meeting used Venice as the focus for discussion of state-of-the-art knowledge and comparison with examples from around the world:

The primary aim of the meeting was to provide a forum for discussion. It brought together over 130 researchers and practitioners from a broad range of disciplines. Participants were invited to illustrate key research results with self-criticism and to suggest realistic and demonstrable ways of solving current problems. Specifically, the conference sought to:

● identify and discuss the key issues facing Venice and proposed interventions and solutions to combat the flooding and environmental problems;
● provide a synthesis of sound scientific and technological research results;
● explore scale (global-to-local) and uncertainty issues associated with predicting the impact of global environmental change in Venice;
● continue to build a community of researchers for issues related to the sustainable management of coastal lagoon systems using Venice as a case study.

This initiative, due to be published in a volume by Cambridge University Press, provides the first international synthesis of the extensive interdisciplinary research investigation on the Venice problem since the UNESCO Report of 1969.

Project team

Dr Pierpaolo Campostrini – Director, CORILA
Lady Clarke – President, Venice in Peril
Jane Da Mosto – Researcher, CORILA
Dr Caroline Fletcher – Venice Fellow, Department of Geography, University of Cambridge
Prof Peter Guthrie – Centre for Sustainable Development, Department of Engineering, University of Cambridge
Paul Richens – Vice Master, Churchill College
Anna Somers Cocks – Chairman, Venice in Peril
Prof Robin Spence – Director, Cambridge University Centre for Risk in the Built Environment
Dr Tom Spencer (Chair) – Director, Cambridge Coastal Research Unit, University of Cambridge

● Cambridge University Coastal Research Unit

http://ccru.geog.cam.ac.uk/

CCRU carries out fundamental research on coastal, estuarine and near-shore processes, landforms and ecosystems; environmental monitoring in the coastal zone, and research consultancies for both governmental and non-governmental agencies. It works in both temperate and tropical environments. The Unit's brief is to:

● Provide scientifically-informed input for the better management of shorelines and their associated ecosystems.
● Facilitate and promote multi-disciplinary research into all aspects of shallow water marine science by bringing together natural and social scientists in Cambridge University and other governmental and non-governmental research institutions.
● Offer scientifically-informed advice on the sustainable management of coasts and coastal ecosystems for coastal management and

decision-making within governmental and non-governmental institutions and organisations in the UK and overseas.

● Churchill College, University of Cambridge
www.chu.cam.ac.uk

The college was endowed by some of the world's most famous companies, charitable foundations and benefactors, which enabled Sir Winston Churchill to launch his remarkable educational venture – Churchill College. The college was created as an international multidisciplinary community for the study of the sciences and the humanities in order to meet the national need for scientists and technologists, and to forge links with industry. It is one of 31 colleges of the University of Cambridge, and is set in 40-acre grounds. As the first major work of modern architecture in the University, it has been judged "an outstanding conception", "the best of the new".

GLOSSARY

Acqua alta Literally 'high water' in Italian, the term is used to describe periodic flooding in Venice city and islands, usually during the winter months. It has been a feature of Venice for centuries but incidents of high water are becoming increasingly common as average relative water level rises.

Algal bloom Sudden, massive growths of microscopic and macroscopic plant life, algae and cyanobacteria, which develop in lakes, reservoirs and marine waters. Algal bloom is usually the result of urban and agricultural runoff from the watershed area and may result in the death of other aquatic life, including fish and seagrasses. It was the main water quality issue for Venice in the 1980s-90s. (See 'eutrophication')

Aquifer A geological formation or structure that stores water, like a giant underground sponge. The water-bearing rocks that compose aquifers consist either of unconsolidated (soil-like) deposits or consolidated rocks. Water flowing into recharge areas - land covered with soil and trees – refills the aquifer; wetlands absorb and store water that later slowly drains into aquifers. To tap the groundwater (qv) in an aquifer, wells are dug until they reach the top layer of the aquifer, the water table. When a lot of water is pumped from an aquifer, or when there is a dry spell, the water table sinks lower and causes land subsidence (q.v.).

Bora Named after *Boreas*, God of the North Winds, a strong, cold and dry north-easterly wind that blows from the Alps across the Adriatic. It can be very dangerous in the lagoon, provoking high waves. Bora winds are most common during the cool season (November through March).

Breakwater A protective structure of stone or concrete; usually protecting a harbour or beach, designed to break the force of the waves and so prevent a beach from washing away or mitigate the exchange of water between the Adriatic and the lagoon. Breakwaters are typically built parallel to the coast.

Biodiversity A contraction of 'biological diversity', to describe the number, variety and variability of living organisms. Biodiversity is commonly defined in terms of the variability of genes, species and ecosystems, corresponding to these three fundamental and related levels of biological organization. It is significant as regards individual habitat types and for the whole lagoon region.

Drainage basin The tract of land that feeds a given body of water (lake, lagoon, etc) with water originating as precipitation (snow, rain). The water may drain through the ground or be carried in rivers and streams. Other terms include 'catchment area' and 'watershed'.

Ecosystem A basic functional unit in ecology: all the organisms in a particular area, and the non-living environment that supports them. All components are considered inseparable and depend on each other, either directly or indirectly. Ecosystems are dynamic: their boundaries and constituent parts change over time.

Endocrine disruptors These are synthetic chemicals that when absorbed into the body either mimic or block hormones, which disrupt the body's normal functions. Many chemicals, particularly pesticides and plasticisers, are suspected endocrine disruptors. They are typically persistent in the environment and they accumulate in fat.

Erosion The wearing away of land, soil or built structure by the action of wind or water or both (waves). In the lagoon, erosion mainly occurs along the edges of the saltmarshes and between the shallows and the inner channels, so that sediment is carried out to sea by the current or it deposits in the channels which then have to be mechanically dredged.

Eustacy (eustasy) Changes in sea level on a global scale due to expansion ocean volume and/or factors relating to global warming. Usually refers to worldwide change – an increase or decrease in sea level due to more precipitation returning to the ocean (which indicates glacial ablation or inter-glaciation). Changes in sea level can result from movement of tectonic plates altering the volume of ocean basins, or when changes in climate affect the volume of water stored in glaciers and polar icecaps. Eustacy affects positions of shorelines.

Eutrophication The process by which excess nutrients (q.v.) entering a body of water leads to the excessive growth of particular aquatic plants, at the expense of other species. The water becomes starved of oxygen, producing an environment that does not readily support aquatic life. The most serious problems of eutrophication result from the massive growth of single-celled algae (see algal bloom).

Gabion Steel wire-mesh basket to hold stones or crushed rock, to protect a bank or bottom from erosion. Now used in saltmarsh restoration experiments in the lagoon, where gabions of more biodegradable materials are also being trialled.

Global warming An increase in the average near-surface temperature of the Earth. Global warming has occurred in the distant past as the result of natural influences, but the term is most often used to refer to the warming predicted to occur as a result of increased emissions of greenhouse gases (carbon dioxide, methane, nitrous oxide and other gases released into the atmosphere by the burning of fossil fuels.) Thought to be responsible for changes in global climate patterns. Scientists generally agree that the Earth's surface has warmed by about 1°F in the past 140 years.

Greenhouse effect A popular term used to describe the roles of water vapour, carbon dioxide, and other trace gases in keeping the Earth's surface warmer than it would be otherwise. These gases trap solar radiation, as they allow incoming sunlight to pass through but absorb heat radiated back from the Earth's surface. Many human activities cause levels of these gases to rise, resulting in an increase in the Earth's temperature.

Groundwater This is water held within the interconnected openings of saturated rock (usually in aquifers, q.v.) beneath the land surface, that can be used to supply wells and springs. The long filtration route means the groundwater is relatively clean; however contamination can reach groundwater reserves and is generally expensive and difficult to clean up.

Hydrodynamics The study of fluids in motion and the movement of objects through fluid. In relation to the subject of this book, how water moves in the lagoon, and the effect of natural and built structures on its flow – from the winding tidal creeks to the breakwaters at coastal inlets.

Habitat The place where a plant or animal species naturally lives and grows, finding the nutrients, water, shelter, living space, and other essentials it needs to survive. Habitat loss, which includes the destruction, degradation, and fragmentation of habitats, is the primary cause of biodiversity loss.

Insulae These are the islets within the historical centre of Venice and the term is now used to refer to a local flood defence strategy which involves raising the perimeter of individual islets and waterproofing measures, pumping systems and valves on the interior. Small-scale movable barriers may also be installed at the entrances of the inner canals connected to the individual islet. This approach has been implemented at Malamocco (Lido) and is underway for the San Marco islet and the island of Burano.

Jetty A protective structure of stone or concrete extending into the sea to influence the current or tide, in order to protect harbours, shores and banks. (1) On open seacoasts, a structure extending into a body of water to direct and confine the tidal flow to a selected channel, or to prevent shoaling. Jetties are built at the entrance to a bay to help deepen and stabilise a channel and facilitate navigation.

Lagoon (coastal/tidal) A shallow area of water separated from the sea by a sandbank or by a strip of low land, and usually connected to the sea by one or more inlets. It is considered to be an open, complex and dynamic system – open in terms of the exchanges with the sea and watershed; complex and dynamic because of the large variations in physical, chemical, ecological and physiological characteristics within it and over time. Exchanges in material and energy are strong across the land/water, air/water and water/sediment interfaces.

Mathematical modelling Computers allow scientists to construct increasingly detailed 'models' of how a system behaves and to ask questions about it. Their effectiveness depends on the prevailing level of understanding of the real world – the main features of the system and the factors influencing it. Models are used to explain past changes, to investigate why observed changes are happening and to predict impacts of planned interventions. Physical processes can be modelled fairly well, but it gets more difficult once biological factors are added. Computer models also depend on accurate data for calibration, validation and prediction. Improved environmental monitoring, instrumentation, computing powers and communications technologies are helping to make models even closer to real systems.

Morphology The scientific study of the forms of things. In biology, the physical form and structure of animals and plants. In geology, the structure, form, and arrangement of rocks and landforms. In the case of a lagoon, its physical form is the result of hydrodynamics (the way water moves and is exchanged) and the influence of plant and animal communities.

Moto ondoso Wave action, which is a key cause of erosion in the lagoon and city. In the lagoon, waves are primarily generated by winds whereas the propeller action of motor boats creates the most destructive wave action within the inner canals.

Mudflat A flat area in the lagoon, covered with a thick layer of mud or sand. Mudflats are usually exposed at low tide but covered at high tide. They may grade into saltmarsh if conditions allow sediment to build up and become stabilised by plants, and the two systems interlink.

Murazzi Sea walls of stone, concrete, or other sturdy material, built along the shoreline to prevent erosion and other damage by wave action. The lagoon's *murazzi* were first built in the 18th century and constitute the first line of flood defence. Parts of the *murazzi* were broken during the 1966 flood.

Nutrients Elements or compounds essential for animal and plant growth. Major plant nutrients are phosphorus and nitrogen. In excess concentrations they can cause imbalances and eutrophication (q.v.).

Saltmarsh A coastal wetland ecosystem that is periodically inundated by seawater. Plants in this community have special adaptations to survive in salty conditions that would kill most land plants. Saltmarshes support vital physical and biological processes that govern the health of the lagoon. Venice's saltmarshes, locally known as *barene*, offer a unique environment within the Mediterranean thanks to the marked tidal cycle and the stabilisation of certain dynamics by human intervention over ten centuries.

Seagrass Aquatic plants with a developed root system anchored underwater in shallow beds. Seagrass beds are important lagoon habitats, especially for young fish, and help to stabilise the lagoon bed.

Sediment There are various classes of sediment, from fine mud and silt to coarser sand, suspended in or settled out of water. In a lagoon, sediment is continually shifting, to be deposited in other areas within the lagoon or out at sea.

Seiches Seiches are periodic oscillations of water in a closed basin that follow a storm surge, rather as bath water continues to rise and fall for a while after you set the water in motion. One or more seiches following a storm surge may cause repeated flooding or expand the duration of a flood episode.

Sirocco A hot, dry, dust-laden wind that blows from the Sahara Desert north or northwest across North Africa. It picks up some moisture as it travels over the Mediterranean. It is associated with storms in the Mediterranean Sea, and cold, wet weather in Europe. The Sirocco's duration may be a half day or many days.

Storm surge A rise in sea level caused by wind and pressure systems; its typical effect is to raise the level of the tide above the predicted (astronomical) level. The magnitude of the storm surge is dependent on the severity and duration of the event and the seabed topography at the site. Storm surges are potentially catastrophic, especially in deltaic regions with onshore winds at the time of high water level and extreme wind wave heights.

Subsidence Decrease in the elevation of land surface due to tectonic, seismic, or artificial forces – such as the loss of underground water support. It is also brought about by land use changes that may directly compact the ground (e.g. heavy buildings) or accelerate oxidation rate of the organic content of soils.

Sustainability The concept of meeting the needs of the present without compromising long-term balance and availability of resources. In nature conservation terms, it refers to the use of a natural resource in a way where it can be renewed, thus maintaining the environment's natural qualities. Today, it applies to many disciplines, including economic development, environment, food production, energy and lifestyle.

Tidal creek A small tidal channel through a coastal marsh, draining into open water and sometimes forming an internal pan within the saltmarsh.

Tides (and how they work) The periodic rise and fall of sea level under the gravitational pull of the moon on the Earth's waters. The sun also exerts a pull, but less than half that of the moon. Especially high tides ('spring tides') occur when the sun and moon are lined up with the Earth at new and full phases of the moon. Conversely, when the sun and moon are on opposite sides of the Earth, they interfere with each other and tides are generally weaker; these are called neap tides. Tides occur regularly, twice every day.

Watershed See drainage basin.

Wetland An area that is regularly saturated by surface water or groundwater and is characterised by vegetation adapted for life in saturated soil conditions (e.g. swamps, bogs, fens, marshes and estuaries); a transitional zone between dry land and aquatic areas, staying wet at least part of the year because the water table is at the surface.

PICTURE CREDITS

INDEX